Synthesis Lectures on Mechanical Engineering

This series publishes short books in mechanical engineering (ME), the engineering branch that combines engineering, physics and mathematics principles with materials science to design, analyze, manufacture, and maintain mechanical systems. It involves the production and usage of heat and mechanical power for the design, production and operation of machines and tools. This series publishes within all areas of ME and follows the ASME technical division categories.

Carlos Alberto Dutra Fraga Filho

Computational Algorithms for Reflective Boundary Conditions in SPH Fluid Dynamics Simulation

Documented Codes

Springer

Carlos Alberto Dutra Fraga Filho
Federal Institute of Education,
Science and Technology
Jucutuquara, Vitória, Espírito Santo, Brazil

The use, adaptation, and reproduction of computational routines are authorised, provided the source is acknowledged.

ISSN 2573-3168 ISSN 2573-3176 (electronic)
Synthesis Lectures on Mechanical Engineering
ISBN 978-3-031-99369-5 ISBN 978-3-031-99370-1 (eBook)
https://doi.org/10.1007/978-3-031-99370-1

This Springer imprint is published by the registered company Springer Nature Switzerland AG
The registered company address is: Gewerbestrasse 11, 6330 Cham, Switzerland

"Thine eyes did see my substance, yet being unperfect; and in thy book all my members were written, which in continuance were fashioned, when as yet there was none of them."

Psalm 139:16 KJV

This book is dedicated to the Lord, who did not abandon me halfway; He extended His hand to me when I was weak and fallen, and saved me from the hands of those who wanted to harm me. He also granted me the health and wisdom to complete this humble work.

Preface

This book aims to introduce the reader to the computational algorithms used in the case studies presented in "Computational Algorithms for Reflective Boundary Conditions in SPH Fluid Dynamics Simulation." The objective is to provide the student with basic routines that could be applied in the simulations presented in the case studies in that book. There are two routines in the two-dimensional domain and one in the three-dimensional domain, which can be used by the researcher.

Far from having the objective of presenting parallelized routines or ready to be used in GPU CUDA programming, we seek to provide researchers/students with basic computational programs that validate the case studies presented in the original literature and that can be adapted by the researcher for the simulation of other cases.

It is considered necessary that the reader has basic knowledge of the fundamentals of the Smoothed Particle Hydrodynamics method, as well as the FORTRAN programming language so that he can make good use of the codes provided.

Carlos Alberto Dutra Fraga Filho

Vitória, Brazil
May 2025

Acknowledgements

The author thanks Springer Nature for the opportunity to publish this book, complementary to the homonymous publication carried out in 2024, containing material used in computational simulations using the Smoothed Particle Hydrodynamics Lagrangian Particle Methods.

Carlos Alberto Dutra Fraga Filho

Contents

About the Author

Carlos Alberto Dutra Fraga Filho is a Brazilian professor and researcher. He holds the following degrees: undergraduate in Mechanical Engineering (1998), Master of Science in Mechanical Engineering (2007), and PhD in Environmental Engineering (2014). He has experience in the mechanical and environmental engineering fields, with an emphasis on Computational Fluid Dynamics and Transport Phenomena. Fraga Filho is a Journals' Reviewer: Physics of Fluids (American Institute of Physics), Journal of the Brazilian Society of Mechanical Sciences and Engineering, and Advances in Water Resources, and has two books—Smoothed Particle Hydrodynamics: Fundamentals and Basic Applications in Continuum Mechanics (https://doi.org/10.1007/978-3-030-00773-7) and Dynamic Analysis of Composite Materials (FGM) via Finite Elements: Numerical Simulation and Experimental Validation (ISBN-10: 6202043490)—and diverse scientific papers published.

Chapter 1
Introduction

This book complements the work of the same name published by this publisher in 2024 [1]. Its objective is to make available to researchers computational routines used in numerical simulation results presented in that publication.

This work's contribution lies in providing the computational routine that presents the reflective boundary conditions, coupled with the computational module dedicated to the Smoothed Particle Hydrodynamics (SPH) method. The implementation began in two-dimensional (2D) domains, and development continued for a few years so that the three-dimensional (3D) implementation was carried out consistently, allowing the validation of some cases in this latter domain.

The computational routines presented here refer to the following cases:

1. Uniform and Incompressible Fluid at Rest Inside a Reservoir (2D case), whose physical-mathematical modelling and numerical results were presented in references [1,2,3]
2. Dam-Breaking Over a Dry Bed (2D case) [1,2,3]
3. Dam-Breaking Over a Dry Bed (3D case) [1,4]

It is relevant to present the following considerations:

- The number of particles, parameters, and initial conditions used in simulations is essential to guarantee the convergence of the results. If the parameters in the routines are changed, the simulation results can also be substantially altered. In this manner, if the reader wants to explore the routines with different combinations of particles and parameters, it will be necessary to do studies of the convergence of the results (which is normal in the SPH method simulations).

- The routines in the two-dimensional domain can be extended to three-dimensional cases through adaptations, such as formulation of the kernel for the 3D domain, adequate initial distribution of the fluid particles in the domain, and adjustments in the dimensions of the matrices for the 3D domain.

C. A. Dutra Fraga Filho, *Computational Algorithms for Reflective Boundary Conditions in SPH Fluid Dynamics Simulation*, Synthesis Lectures on Mechanical Engineering,
https://doi.org/10.1007/978-3-031-99370-1_1

- Reference [5] presents the Corrective Smoothed Particle Method (CSPM), applied in the restoration of the consistency in the regions close to the boundaries
- References [6-7] present the formulation of artificial pressure, which can be used in the simulation to avoid the tensile instability in the SPH method (and which is implemented in the last program presented in this work)
- Additional references [8-10] are also cited, considering that the case studies presented therein contributed to developing the computational routines and may be useful for the reader.

Finally, directories containing the computational routines are available as supplementary material in each of the chapters of this book. Simple scripts in MATLAB are provided for plotting simple graphs showing the spatial evolution of particles in time.

References

1. Fraga Filho, C.A.D. Reflective Boundary Conditions in SPH Fluid Dynamics Simulation Two and Three-dimensional Validation and Applications. Synthesis Lectures on Mechanical Engineering, Springer Nature, 2024

2. Fraga Filho, C.A.D. Smoothed Particle Hydrodynamics. Fundamentals and Basic Applications in Continuum Mechanics. Springer Nature, 2019

3. Fraga Filho, C.A.D. An algorithmic implementation of physical reflective boundary conditions in particle methods: Collision detection and response. Physics of Fluids 29, 113602, 2017. https://doi.org/10.1063/1.4997054

4. Fraga Filho C.A.D., Peng C., Islam Md R.I., McCabe C., B. Samiullah, Prasad G.V.D. Implementation of three-dimensional physical reflective boundary conditions in mesh-free particle methods for continuum fluid dynamics: Validation tests and case studies. Physics of Fluids 31, 103606, 2019. https://doi.org/10.1063/1.5115776

5. Chen, J.K., Beraun, J.E., Carney, T.C.: A corrective smoothed patricle method for boundary value problems in heat conduction. Int. J. Numer. Methods Eng. 46, 231–252, 1999.

6. Monaghan, J.J. SPH without tensile instability. J. Comput. Phys. 159, 290–311, 2000.

7. Lobovský, L., Kren, J. Smoothed particle hydrodynamics modelling of fluids and solids. Applied and Computational Mechanics, 2007.

8. Fraga Filho C.A.D. Development of a computational instrument using a lagrangian particle method for physics teaching in the areas of fluid dynamics and transport phenomena. https://doi.org/10.1590/1806-9126-RBEF-2016-0289

9. Fraga Filho C.A.D., Schuina L.L., Porto B.S. An Investigation into Neighbouring Search Techniques in Meshfree Particle Methods: An Evaluation of the Neighbour Lists and the Direct Search. Arch Computat Methods Eng 27, 1093–1107 (2020). https://doi.org/10.1007/s11831-019-09345-9

10. Fraga Filho, C.A.D, Piccoli, F.P. Diffusive terms applied in smoothed particle hydrodynamics simulations of incompressible and isothermal Newtonian fluid flows. J Braz. Soc. Mech. Sci. Eng. 43, 479, 2021. https://doi.org/10.1007/s40430-021-03158-3

Chapter 2

Uniform and Incompressible Fluid at Rest Inside a Reservoir

```
program main

! A file named "main.f90" can be created

!---------------------------------------------------------------------------------------------------
! mass -- constant mass of particles                                    [in]
! ntotal -- total particle number                                       [in]
! dt -- Time step                                                       [in]
! x -- coordinates of particles                                         [in/out]
! rho -- densities of particles                                         [in/out]
! hsml -- smoothing lengths of particles                                [in/out]
!---------------------------------------------------------------------------------------------------

   implicit none
   include 'param.inc'

   integer, dimension (:,:), allocatable :: neighbours
   double precision, dimension (:), allocatable :: rho, mass, p, viscosity, hsml,
sum_CSPM
   double precision, dimension (:,:), allocatable :: x, v, delta_x, w, r, dist, dwdx,
dwdy,dx, dy, pressure_force, viscous_force, gravity, accel, x_initial, n_g, X_0
   integer nx, ny, ntotal, itimestep, maxtimestep, d, m, n, mp, np, i, hours, minutes,
hours_in_seconds, cont_domain, change
   double precision start, finish, time, seconds, xlmin, xlmax, ylmin, ylmax, dx1, s1, s2,
dt, K_scale, factor, kh, radius, CR, H
   character(len=3) aux
   character(len=3) equilibrium

   call cpu_time(start)

   equilibrium ='YES'  ! Assuming initial hydrostatic equilibrium
   itimestep=1

!---------------------------------------------------------------------------
!-------------------------USER DEFINITIONS----------------------------
!---------------------------------------------------------------------------

!----------- Defining the computational domain --------------------
   xlmin=0.0
   xlmax=1.0
   ylmin=xlmin
   ylmax=xlmax
```

Supplementary Information The online version contains supplementary material available at https://doi.org/10.1007/978-3-031-99370-1_2.

C. A. Dutra Fraga Filho, *Computational Algorithms for Reflective Boundary Conditions in SPH Fluid Dynamics Simulation*, Synthesis Lectures on Mechanical Engineering, https://doi.org/10.1007/978-3-031-99370-1_2

```
!------------ Defining the number of particles ----------
    nx=10  ! Defining the number of cells in the x-direction
    ny=nx  ! Defining the number of cells in the y-direction

    m=5    ! Defining the number of particles on each cell direction (x-direction)/ 5 =
default
    n=m    ! Defining the number of particles in y-direction)

    mp=m*nx   ! Number of particles per domain line
    np=n*ny   ! Number of particles per domain column

    ntotal=mp*np      ! Total number of particles

    dt=1.e-4      ! Defining the simulation time step

    CR = 1.00 ! Defining the coefficient of restitution of kinetic energy (reflective
boundary conditions)

    maxtimestep = 10000   ! Physical time simulated of 1.00 s

    print*, 'The total number of particles for this simulation:'
    print*, mp, 'x', np
    print*, ''
    print*, 'coefficient of restitution of kinetic energy = ' , CR

    allocate (neighbours(maxn,n_neighbours), rho(maxn), hsml(maxn), p(maxn),
viscosity(maxn), w(maxn, n_neighbours), dwdx(maxn, n_neighbours), dwdy(maxn,
n_neighbours), x(maxn,2), v(maxn,2), r(maxn, n_neighbours), pressure_force(maxn,2),
viscous_force(maxn,2), gravity(maxn,2), dist(maxn, n_neighbours), dx(maxn,
n_neighbours), dy(maxn, n_neighbours), mass(maxn), accel(maxn,2),
sum_CSPM(maxn), n_g(4,2), X_0(4,2))

    open(unit=445,file='../output/normals.dat')
    open(unit=450,file='../output/flat_points.dat')
    open(unit=455,file='../output/positions_and_velocities_of_centres_of_mass.dat')
    open(unit=66,file="../output/collision_history.dat")
    open(unit=67,file="../output/collision_coordinates.dat")
    open(unit=68,file = '../output/differences_in_positions.dat')
    open(unit=69,file= '../output/particle_movement.dat')

    call geometry (xlmax,ylmax, n_g, X_0)

    call input(itimestep, xlmin, xlmax, ylmin, ylmax, m, mp, np,  nx, dx1, kh, mass, rho,
hsml, K_scale, viscosity, cont_domain, x,v, H)

    radius = dx1/2.  ! Particle radius

    open(17,file='../output/matlab_parameters.dat')
    rewind(17)
    write(17,18) xlmin, xlmax, ylmin, ylmax
    close(17)
```

```
18   format (4(1x,e10.3))

    print*, ' '
    print*, '-------------------------------------------------------------------------------'
    print*, '------ WELCOME TO THE START OF THE SIMULATION ------'
    print*, '-------------------------------------------------------------------------------'
    print*, ' '
    print*, ' '
    print*, 'There will be', maxtimestep, ' numerical iterations', ' using the Kernel' , skf ,
'(See param.inc to identify the Kernel)'
    print*,' '
    print*, 'domain length (m) = ', xlmax-xlmin
    print*,' '
    print*, 'domain heigth (m) = ', ylmax-ylmin
    print*,' '
    print*, 'number of cells in x-direction = ', nx , ' -- number of cells in y-direction = ',
ny
    print*,' '
    print*, 'number of particles per side of the cell = ', m
    print*,' '
    print*, 'total number of particles in the domain = ', m*nx*n*ny
    print*,' '
    print*, 'time step (dt, em seconds) = ', dt
    print*,' '
    print*, 'Press any key to continue... '
    pause

    do i=1,maxtimestep

      call direct_find(itimestep,cont_domain,hsml, neighbours,x,r,dx,dy,K_scale)

       if (equilibrium.eq.'YES') then
         call modified_pressure(itimestep, p, cont_domain)  ! hydrostatic equilibrium
         print*,  ' '
         print*,  ' ---- Hydrostatic equilibrium ---- '
         print*,  ' '
       else
         call absolute_pressure(p, x, rho, cont_domain, H)  ! particle motion detected
         print*,  ' '
         print*, ' ---- Particle motion detected ---- '
         print*,  ' '
       end if

       call kernel(cont_domain, r, hsml, mass, rho, w,neighbours, dx, dy, dwdx, dwdy,
sum_CSPM,factor)

       call                    internal_force(itimestep,cont_domain,hsml,mass,v,neighbours,r,
pressure_force, viscous_force, rho, viscosity, dwdx, dwdy, p)
       call external_force(cont_domain,gravity)
```

```
    call acceleration(cont_domain, pressure_force, viscous_force, accel, gravity,
equilibrium)

    call time_integration(itimestep,cont_domain,accel, dt, x, v, equilibrium, H)

    call collision_test(cont_domain,radius,itimestep,dt,x,v, n_g, X_0, CR, change)

    if (change.eq.0)  equilibrium ='YES'   ! Hydrostatic equilibrium maintained after
application of reflective boundary conditions

    call output(cont_domain, x, v, accel, p, itimestep)

    itimestep=itimestep+1

    print*, ' '
    print*, '--------------------------------------------------------------------------------'
    print*, ' LEAVING THE ITERATION NUMBER ', itimestep, ' OUT OF A
TOTAL OF ' , maxtimestep
    print*, '--------------------------------------------------------------------------------'
    print*, ' '
    print*, ' Simulated physical time (s) = ', i*dt
    print*, ' '

  end do

  call cpu_time(finish)

  time=finish-start
  hours_in_seconds=3600;   !hours em seconds
  hours = (time/hours_in_seconds)
  minutes = (time -(hours_in_seconds*hours))/60
  seconds = (time -(hours_in_seconds*hours)-(minutes*60))

  print*, ' ------------------------------------------------------------------'
  print*, '                END OF SIMULATION '
  print*, hours ,' hours', minutes,' minutes', seconds, ' seconds.'
  print*, ' ------------------------------------------------------------------'

  open(776,file='simulation_parameters.dat')
  rewind(776)
  write(776,*) ' '
  write(776,*)  'domain length (m) = ',  xlmax-xlmin
  write(776,*) ' '
  write(776,*)  'domain height (m) = ',  ylmax-ylmin
  write(776,*) ' '
  write(776,*)  'number of cells in x-direction = ', nx, ' number of cells in y-direction =
', ny
  write(776,*) ' '
  write(776,*)  'number of particles per side of the cell = ', m
  write(776,*) ' '
```

```
    write(776,*)  ' total number of particles in the domain = ',  m*nx*n*ny
    write(776,*) ''
    write(776,*)  'dx (m) = ', dx1
    write(776,*) ''
    write(776,*)  'time step (dt, em seconds) = ', dt
    write(776,*) ''
    write(776,*)  ' Time for the physical process =', itimestep*dt
    write(776,*) ''
    write(776,*)  'Simulation time = ', hours ,' hours', minutes,' minutes', seconds, '
 seconds.'

    close(91)
    close(445)
    close(450)
    close(455)
    close(66)
    close(67)
    close(68)
    close(69)
    close(64)
    close(776)

 end program main


!---------------------------------------------------------------------------------------------
! Subroutines:
! Files should be created with the name "subroutine_name.f90", for example,
! geometry.f90
!---------------------------------------------------------------------------------------------

subroutine geometry (xlmax,ylmax, n_g, X_0)

!--- In this subroutine the planes that define the walls of the reservoir are defined --------

   implicit none

   integer  i,j,active_planes(4)
   double precision  xlmax, ylmax, n_g(4,2), X_0(4,2),summation

! Defining normal vectors to the planes
! Note: if you want to disable a plane, just define the normal vector to !the plane as
n_0=(0.,0.)

   rewind(445)
   rewind(450)

! The normal vector to the right lateral plane is (-1.,0.) - **plane 01**
   n_g(1,1)=-1.
   n_g(1,2)=0.
```

```
! The normal vector to the left lateral plane is (1.,0.) - **plane 02**
   n_g(2,1)=1.
   n_g(2,2)=0.
! The normal vector to the lower plane is (0.,1.) - **plane 03**
   n_g(3,1)=0.
   n_g(3,2)=1.
! The normal vector to the upper plane is (0.,-1.) - **plane 04**  - DISABLED!
   n_g(4,1)=0.
   n_g(4,2)=0.

! Defining a point belonging to each plane
! Note: here you define the geometry of the reservoir

! Point belonging to the right wall
   X_0(1,1)=xlmax
   X_0(1,2)=ylmax

! Point belonging to the left wall
   X_0(2,1)=0.0
   X_0(2,2)=0.0

! Point belonging to the bottom
   X_0(3,1)=xlmax/2.
   X_0(3,2)=0.

! Point belonging to the upper plane
   X_0(4,1)=0.
   X_0(4,2)=0.

   do i=1,4
    write(445,13) (n_g(i,j),j=1,2)
    write(450,13) (X_0(i,j),j=1,2)
13  format(2(2x,F10.3))
   end do

   do i=1,4
    summation=0.
    do j=1,2
     summation = summation + n_g(i,j)
    end do
    active_planes(i)=summation   ! If plane "i" is active, the value of its position in the
vector "active_planes" will be different from zero
   end do

! Checking active planes

   print*, '------------ Active planes ----------------'
   print*, ' '
   do i=1,4
```

```
    if (active_planes(i).ne.0.) then

     if (i.eq.1) then
       print*, '[1] - [RIGHT SIDE]'
       print*, ' '
     else
       if (i.eq.2) then
         print*, '[2] - [LEFT SIDE]'
         print*, ' '
       else
         if (i.eq.3) then
           print*, '[3] - [BOTTOM]'
           print*, ' '
         else
           if (i.eq.4) then
             print*, '[4] - [UPPER PLANE]'
             print*, ' '
           end if
         end if
       end if
     end if

    end if

  end do

end subroutine geometry


subroutine input (itimestep, xlmin, xlmax, ylmin, ylmax, m, mp, np,  nx, dx1, kh, mass,
rho, hsml, K_scale, viscosity, cont_domain, x,v, H)

!------------------------------------------------------------------------------------------
!   x -- coordinates of particles                                           [out]
!   mass -- mass of particles                                               [out]
!   rho -- densities of particles                                           [out]
!   hsml -- smoothing lengths of particles                                  [out]
!   p -- presuure                                                           [out]
!------------------------------------------------------------------------------------------

   implicit none
   include 'param.inc'

   integer cont_domain
   double precision  xlmin, xlmax, ylmin, ylmax, dx1, dy1, kh, K_scale, rho0, g, b,
gamma
   integer i, j, k, d, nx, ny, m, n ,mp, np,itimestep
   double precision  mass(maxn), rho(maxn), hsml(maxn), viscosity(maxn), aux_H, H
   double precision  x(maxn,2), v(maxn,2)
```

```
   dx1 = ((xlmax-xlmin)/nx)/m     ! Distance between the centre of mass of two adjacent
particles in the x-direction
   dy1 = dx1                      ! Distance between the centre of mass of two adjacent
particles in the y-direction
   kh=((xlmax-xlmin)/nx)/2.       ! Cell radius


!------ SCALE FACTOR FOR DIFFERENT KERNELS USED IN SIMULATION ------

   if (skf.eq.1) then  ! Cubic spline kernel
    K_scale=2.0
   else
    if (skf.eq.2) then   ! Quintic spline kernel
     K_scale=3.0
    else
     if (skf.eq.3) then  ! Quartic kernel
      K_scale=1.0
     end if
    end if
   end if

!---------------------- Domain Particle Distribution and Initial Properties --------------------

   open(76, file='../output/data/position/0000000.dat')
   open(77, file='../output/data/velocity/0000000.dat')

   k=0

   do i = 1, mp
     do j = 1, np
      k = k+1
      x(k,1) =  xlmin +(i-1)*dx1 + dx1/2.
      x(k,2) =  ylmin +(j-1)*dy1 + dy1/2.
      v(k,1) = 0.
      v(k,2) = 0.
    end do
  end do

  cont_domain=0

  do i=1,k
    cont_domain=cont_domain+1
    rho (cont_domain) =  1000.
    mass(cont_domain) = dx1*dy1*rho(cont_domain)
    viscosity(cont_domain)=1.0e-3
    hsml(cont_domain) = kh/K_scale

    write(76,225) i, x(i,1), x(i,2)
    write(77,225) i, v(i,1), v(i,2)
  end do
```

```
225 format (I8, 2(1x,D21.14))


!------ Finding H (water height in the reservoir) ----

   aux_H = 0.

   do i=1,cont_domain

     if (x(i,2).gt.aux_H)    aux_H = x(i,2)

   end do

    H = aux_H    !-----------THIS IS THE WATER LEVEL IN THE RESERVOIR AT
EACH NUMERICAL ITERATION

    open(91,file="../output/water_level.dat")
    write(91,*) 'itimestep          H '

    print*,' **************************************************'
    print*,'    NUMBER OF PARTICLES IN THE DOMAIN '
    print*,'    Total number of particles = ', k
    print*,' **************************************************'

end subroutine input


subroutine direct_find(itimestep,cont_domain,hsml, neighbours,x,r,dx,dy,K_scale)

!-----------------------------------------------------------------------------------------------
!   Subroutine to detect and store neighboring particles.
!   Interaction pairs are determined by directly comparing the particles distance with
!   the corresponding support radius.
!
!   cont_domain -- number of particles in the domain                    [in]
!   hsml  -- smoothing length                                           [in]
!   x  -- coordinates of all particles                                  [in]
!   niac  -- number of neighbours                                       [out]
!   distance -- distance between two particles                          [out]
!-----------------------------------------------------------------------------------------------

    implicit none
    include 'param.inc'

    integer  i, j, k, niac, cont_domain, itimestep, neighbours(maxn,n_neighbours)
    double precision distance, aux, K_scale, support_radius(maxn), hsml(maxn)
    double precision   r(maxn, n_neighbours), dx(maxn, n_neighbours), dy(maxn,
n_neighbours), x(maxn,2)
    do i=1,cont_domain
      support_radius(i) = K_scale*hsml(i)
```

```
    end do

    do i=1,cont_domain
      do j=1, n_neighbours
        neighbours(i,j)=0
        dx(i,j)=0.
        dy(i,j)=0.
        r(i,j)=0.
      end do
    end do

    do i=1, cont_domain
      niac= 2
      neighbours(i,1)=i

      do j = 1, cont_domain
        aux  = (x(j,1) - x (i,1))**2 + (x(j,2) - x(i,2))**2
        distance = sqrt(aux)

        if (distance.le. support_radius(i)) then
          neighbours(i,niac)=j
          r(i,niac)=distance
          dx(i,niac)=(x(j,1) - x(i,1))
          dy(i,niac)=(x(j,2) - x(i,2))
          niac=niac+1
        end if

      end do

    end do

    open(222,file='../output/neighbours.dat')  ! Storing the neighbours of each particle in
the domain
    do i=1,cont_domain
      write(222,*) (neighbours(i,j), j=1, n_neighbours)
    end do
    close(222)

end subroutine direct_find


subroutine modified_pressure(itimestep, p, cont_domain)

!----------------------------------------------------------------------------------------------------
!   cont_domain --  number of particles in the domain                          [in]
!   rho -- density                                                             [in]
!   p    : pressure                                                            [out]
!   H    : water level in the reservoir                                        [out]
!----------------------------------------------------------------------------------------------------
```

```
   implicit none
   include 'param.inc'

   integer  i, cont_domain, itimestep
   double precision  p_mod(maxn), p(maxn)

open(78, file='../output/data/pressure/0000000.dat')

 !------- Calculating modified pressure for all particles at the domain

   do i=1,cont_domain
     p_mod(i) = 0.      ! The modified pressure is a constant (in hydrostatic equilibrium)
     p(i) = p_mod(i)
   end do

   if (itimestep.eq.1)  then
    open(78, file='../output/data/pressure/0000000.dat')
    do i=1,cont_domain
      write (78,222) i, p(i)
    end do
   end if

222 format (I8, 1x, D21.14)

end subroutine modified_pressure


subroutine absolute_pressure (p, x, rho, cont_domain, H)

!----------------------------------------------------------------------------------------------------
!   cont_domain --  number of particles in the domain                                     [in]
!   rho  -- density                                                                       [in]
!   p -- pressure                                                                         [out]
!   c  -- sound velocity                                                                  [out]
!  H -- water level in the reservoir                                                      [in]
!----------------------------------------------------------------------------------------------------

   implicit none
   include 'param.inc'

   integer  n_part, b, cont_domain
   double precision rho(maxn), p(maxn), x(maxn,2)
   double precision gamma, rho0, grav, H

!  Parameters and constants used in Tait's Equation
   gamma=7.
   rho0=1000.
   b = 1.0e5
   grav = 9.81        ! magnitude of gravity
```

```
    do n_part=1,cont_domain
      p(n_part) = ( b*((rho(n_part)/rho0)**gamma-1)  ) + ( rho(n_part)*grav*(H-x(n_part,2) ) )      ! dynamic pressure + pressure exerted by the water column on the particle
    end do

    print*, '------------------------------------------------------------------------'
    print*, '-------------------- There was particle motion. --------------------'
    print*, '------------------------------------------------------------------------'

  end subroutine absolute_pressure


subroutine kernel (cont_domain, r, hsml ,mass, rho, w, neighbours, dx, dy, dwdx, dwdy, sum_CSPM, factor)

!-------------------------------------------------------------------------------------------
!  Subroutine to calculate the smoothing kernel and its derivatives
!
!  if skf = 1, Cubic spline kernel   (defined in param.inc)
!          = 2, Quintic kernel
!          = 3, Quartic kernel by Liu and Lam
!  cont_domain -- number of prticles in the domain                       [in]
!  r -- distance between particles i and j                               [in]
!  dx -- x-distance between i and j (fixed and neighbour particle)       [in]
!  dy -- y-distance between i and j (fixed and neighbour particle)       [in]
!  hsml  -- smoothing length                                             [in]
!  w -- kernel for all interaction pairs                                 [out]
!  fator -- kernel's normalisation constant                              [out]
!  dwdx -- derivative of kernel with respect to x-direcion               [out]
!  dwdy -- derivative of kernel with respect to y-direction              [out]
!-------------------------------------------------------------------------------------------

    implicit none
    include 'param.inc'

    integer i, j, niac, neighbours(maxn,n_neighbours), cont_domain
    double precision r(maxn, n_neighbours), w(maxn, n_neighbours), dx(maxn, n_neighbours), dy(maxn, n_neighbours), hsml(maxn), dwdx(maxn, n_neighbours), dwdy(maxn, n_neighbours), mass(maxn), rho(maxn), sum_CSPM(maxn)
    double precision q, factor, aux

    open(88,file='../output/kernel_values.dat')
    rewind(88)
    do i=1, cont_domain
      niac=2
      w(i,1)=0.
      dwdx(i,1)=0.
      dwdy(i,1)=0.
      sum_CSPM(i)=0.        ! To make the CSPM Correction
```

```
do while (neighbours(i,niac).ne.0)
 j=neighbours(i,niac)
 q = r(i,niac)/hsml(i)

 if (skf.eq.1) then     ! Cubic spline kernel

  factor = 15./(7.*pi*hsml(i)*hsml(i))

   if (q.ge.0.and.q.le.1.) then
    w(i,niac)= factor* ((2./3.) - q**2. + (1./2.)*q**3)
    dwdx(i,niac) = factor * ((-2.)/hsml(i)**2 + (3./2.)*(r(i,niac)/hsml(i)**3)) * dx(i,niac)
    dwdy(i,niac) = factor * ((-2.)/hsml(i)**2 + (3./2.)*(r(i,niac)/hsml(i)**3)) * dy(i,niac)

    sum_CSPM(i)= sum_CSPM(i) + (mass(j)/rho(j))*w(i,niac)

    write(88,12) q, w(i,niac)

   else
    if (q.gt.1..and.q.le.2.) then
     w(i,niac) = factor * 1.e0/6.e0 * (2.-q)**3
     dwdx(i,niac) = -factor * (1./(2.*hsml(i))) * ( ( 2. - r(i,niac)/hsml(i))**2. ) * dx(i,niac)/r(i,niac)
     dwdy(i,niac) = -factor * (1./(2.*hsml(i))) * ( ( 2. - r(i,niac)/hsml(i))**2. ) * dy(i,niac)/r(i,niac)

     sum_CSPM(i)= sum_CSPM(i) + (mass(j)/rho(j))*w(i,niac)

     write(88,12) q, w(i,niac)
    end if

   end if

 end if  ! Cubic spline kernel


 if (skf.eq.2) then              ! Quintic spline kernel

  factor = 7. / (478.*pi*hsml(i)*hsml(i))

  if (q.ge.0..and.q.le.1.) then
   w(i,niac) = factor * ( (3.-q)**5. - 6.*(2.-q)**5. + 15.*(1.-q)**5. )
   aux= (-5./hsml(i))* (    (3.- (r(i,niac)/hsml(i))**4 ) - ( 6 * (2.- (r(i,niac)/hsml(i))**4 ) ) ) + ( 15 * (1.- (r(i,niac)/hsml(i))**4 ) )
  dwdx(i,niac) = factor * aux * dx(i,niac)/r(i,niac)
  dwdy(i,niac) = factor * aux * dy(i,niac)/r(i,niac)
  sum_CSPM(i)= sum_CSPM(i) + (mass(j)/rho(j))*w(i,niac)
```

```
      write(88,12) q, w(i,niac)

    end if

    if (q.gt.1..and.q.le.2.) then
      w(i,niac) = factor * ( (3.-q)**5. - 6.*(2.-q)**5. )
      aux = (-5./hsml(i))* ( (3.- (r(i,niac)/hsml(i))**4 ) - ( 6 * (2.-
(r(i,niac)/hsml(i))**4 ) ) )
      dwdx(i,niac) = factor * aux * dx(i,niac)/r(i,niac)
      dwdy(i,niac) = factor * aux * dy(i,niac)/r(i,niac)

      sum_CSPM(i)= sum_CSPM(i) + (mass(j)/rho(j))*w(i,niac)

      write(88,12) q, w(i,niac)

    end if

    if (q.gt.2..and.q.le.3.) then
      w(i,niac) = factor * (3.-q)**5.
      aux = (-5./hsml(i))* (3.- (r(i,niac)/hsml(i))**4 )
      dwdx(i,niac) = factor * aux * dx(i,niac)/r(i,niac)
      dwdy(i,niac) = factor * aux * dy(i,niac)/r(i,niac)

      sum_CSPM(i)= sum_CSPM(i) + (mass(j)/rho(j))*w(i,niac)

      write(88,12) q, w(i,niac)

    end if

  end if   ! quintic spline kernel

  if (skf.eq.3) then       ! Quartic kernel

    factor = 15. / (7.*pi*hsml(i)*hsml(i))

    if (q.ge.0..and.q.le.2.) then
      w(i,niac) = factor * ( (2./3.- (9./8.)*q**2.) + (19./24.)*q**3. - (5./32.)*q**4. )
      aux = ( -18./(8.*hsml(i)**2.) + (57.*r(i,niac))/(24.*hsml(i)**3.) -
(20.*r(i,niac)**2.)/(32.*hsml(i)**4.) )
      dwdx(i,niac) = factor * aux * dx(i,niac)
      dwdy(i,niac) = factor * aux * dy(i,niac)

      sum_CSPM(i)= sum_CSPM(i) + (mass(j)/rho(j))*w(i,niac)

      write(88,12) q, w(i,niac)

    end if
  end if   ! Quartic kernel
```

```
        niac=niac+1

      end do

   end do

12 format(2(1x,D21.14))

close(88)

end subroutine kernel


subroutine internal_force(itimestep, cont_domain, hsml, mass, v, neighbours, r,
pressure_force, viscous_force, rho, viscosity, dwdx, dwdy, p)

!------------------------------------------------------------------------------------------------
! This subroutine calculates the internal forces in the Navier-Stokes equations, including
! the pressure gradient and viscous forces, used by the time integration method
!
! itimestep -- current timestep number                                          [in]
! cont_domain -- number of particles in the domain                              [in]
! hsml -- smoothing length                                                      [in]
! mass -- particle masses                                                       [in]
! niac -- number of interaction pairs                                           [in]
! rho -- density                                                                [in]
! viscosity -- dynamic viscosity                                                [in]
! dwdx -- derivative of kernel with respect to x-direcion                       [in]
! dwdy -- derivative of kernel with respect to y-direcion                       [in]
!------------------------------------------------------------------------------------------------

   implicit none
   include 'param.inc'

   integer cont_domain, itimestep, ntotal,niac, i, j, k,
neighbours(cont_domain,n_neighbours)
   double precision hsml(cont_domain), mass(cont_domain),
v(cont_domain,2),rho(cont_domain), dwdx(cont_domain, n_neighbours),
dwdy(cont_domain, n_neighbours), dwdr(cont_domain, n_neighbours),
r(cont_domain, n_neighbours), p(cont_domain), viscous_force(cont_domain,2),
pressure_force(cont_domain,2), viscosity(cont_domain)
   double precision q, aux, aux_2, aux_3

!------------------------------------------------------------------------------------------------
!---------------- CALCULATION OF LAMINAR VISCOUS FORCES -------------------
!------------------------------------------------------------------------------------------------
    do k=1,cont_domain
     niac=2
     i = neighbours(k,1)
     viscous_force(i,1)=0.
```

```
    viscous_force(i,2)=0.

      do while (neighbours(i,niac).ne.0)
        j = neighbours(i,niac)
        aux = (viscosity(j)+viscosity(i)) * (mass(j)/ rho(j))
        aux_2 = r(i,niac)*dwdr(i,niac) * (v(j,1)-v(i,1))
        aux_3 = r(i,niac)*dwdr(i,niac) * (v(j,2)-v(i,2))
        viscous_force(i,1)=viscous_force(i,1) + aux * aux_2
        viscous_force(i,2)=viscous_force(i,2) + aux * aux_3
        niac=niac+1
      end do
  end do

!-----------------------------------------------------------------------------------
!------------------------- CALCULATION OF PRESSURE FORCES -------------------------
!-----------------------------------------------------------------------------------

   do k=1, cont_domain
     niac=2
     i = neighbours(k,1)
     pressure_force(i,1)=0.
     pressure_force(i,2)=0.

     do while (neighbours(i,niac).ne.0)

       j = neighbours(i,niac)

       if (pa_sph.eq.1) then

         aux_2 =  -  ( 1./(rho(i)) ) * ( p(j) -p(i) )* dwdx(i,niac) * mass(j)    ! in Cartesian
coordinates
         aux_3 =  -  ( 1./(rho(i)) ) * ( p(j) -p(i) )* dwdy(i,niac) * mass(j)

       else
         if (pa_sph.eq.2) then

           aux_2 =  - ( (p(i)/rho(i)**2.) + (p(j)/rho(j)**2) ) * dwdx(i,niac) * mass(j)   ! in
Cartesian coordinates
           aux_3 =  - ( (p(i)/rho(i)**2.) + (p(j)/rho(j)**2) ) * dwdy(i,niac) * mass(j)
         end if
       end if

       pressure_force(i,1)=pressure_force(i,1) +  aux_2
       pressure_force(i,2)=pressure_force(i,2) +  aux_3

       niac=niac+1

     end do

   end do
```

```
end subroutine internal_force


subroutine external_force (cont_domain,gravity)

!---------------------------------------------------------------------------------------------------
!   Subroutine to calculate the external forces, in this case, the gravitational forces
!---------------------------------------------------------------------------------------------------

    implicit none
    include 'param.inc'

    integer i, cont_domain
    double precision  gravity(maxn,2)

    do i = 1, cont_domain
      gravity(i,1)=0.
      gravity(i,2) = -9.81
    end do

end subroutine external_force


subroutine acceleration (cont_domain, pressure_force, viscous_force, accel, gravity,
equilibrium)

!---------------------------------------------------------------------------------------------------
!   Subroutine to calculate the particle acceleration
!---------------------------------------------------------------------------------------------------

    implicit none
    include 'param.inc'

    integer n_part, cont_domain
    double precision pressure_force(maxn,2), viscous_force(maxn,2), gravity(maxn,2),
accel(maxn,2)
    character(len=3) equilibrium

    if (equilibrium.eq.'YES') then

      do n_part=1, cont_domain   ! Using the concept of modified pressure in the Navier-
Stokes equation
        accel(n_part,1) = -pressure_force(n_part,1) + viscous_force(n_part,1)
        accel(n_part,2) = -pressure_force(n_part,2) + viscous_force(n_part,2)
      end do

    else
      do n_part=1, cont_domain    ! Using the classic form of the Navier-Stokes equation
```

```
      accel(n_part,1)  =  -pressure_force(n_part,1)  +  viscous_force(n_part,1)  +
gravity(n_part,1)
      accel(n_part,2)  =  -pressure_force(n_part,2)  +  viscous_force(n_part,2)  +
gravity(n_part,2)
     end do

   end if

end  subroutine acceleration


subroutine time_integration(itimestep,cont_domain,accel, dt, x, v, equilibrium, H)

!-----------------------------------------------------------------------------------------
! itimestep -- current timestep number                                             [in]
! x  -- coordinates of particles                                              [in/out]
! v  -- velocities of particles                                               [in/out]
! dt -- timestep                                                                   [in]
! accel -- acceleration of particles                                               [in]
!-----------------------------------------------------------------------------------------

   implicit none
   include 'param.inc'

   integer itimestep, i, k, cont_domain
   double precision x(maxn,2), v(maxn,2), accel(maxn,2), dt, aux_H, H
   character(len=1) equilibrium

! --- Files used to store particle positions and velocities at the beginning of the numerical
iteration  (before temporal integration)

   open(unit=40,file='../output/initial_velocity.dat')    !particles velocities
   open(unit=43,file='../output/initial_position.dat')    !particles positions
   rewind(40)
   rewind(43)

   do i=1, cont_domain
    write(43,31)  (x(i,k), k=1,2)
    write(40,31)  (v(i,k), k=1,2)
   end do
 31  format(2(1x,D21.14))

 close(40)
 close(43)

! --- Performing temporal integration
!    Euler integration method

   do i=1, cont_domain
```

```
      v(i,1) = v(i,1) + accel(i,1)*dt
      v(i,2) = v(i,2) + accel(i,2)*dt
      x(i,1) = x(i,1) + v(i,1)*dt
      x(i,2) = x(i,2) + v(i,2)*dt

      if ( (v(i,1).ne.0.).or.(v(i,2).ne.0.) ) then
        equilibrium = 'NO'
      end if
    end do

! ----- Water level in the reservoir

    aux_H =0.

    do i=1, cont_domain

      if (x(i,2).gt.aux_H)    aux_H = x(i,2)

    end do

    H = aux_H

    write(91,33) itimestep, H
 33   format(I8, D21.4)

end subroutine time_integration


subroutine collision_test(cont_domain,radius,itimestep,dt,x,v, n_g, X_0, CR, change)

!---------------------------------------------------------------------------------------------
! In this subroutine, particle collisions against the boundaries are detected and reflections
! towards the interior of the domain are performed.
!
!  cont_domain --  number of particles in the domain                                    [in]
!  radius --   particles' radius                                                        [in]
!  itimestep – number of the numerical iteration                                        [in]
!  dt --  timestep                                                                      [in]
!  x --  coordinates of particles                                                   [in/out]
!  v --  velocities of particles                                                    [in/out]
!  n_g -- normal vector to plane                                                        [in]
!  X_0  --  point belonging to the plane                                                [in]
!  CR  --  coefficient of restitution of kinetic energy                                 [in]
!  change -- indication of the change in the particle's position in the iteration     [out]
!---------------------------------------------------------------------------------------------

    implicit none
    include 'param.inc'
    integer itimestep,change,cont_domain
```

```
   integer n_part, i, j, aux, ncol, plane_col, plane_col2, plane, cont, d
   double precision  radius,dt, CR
   double precision  aux_x, aux_y, dx, dy, module_a, num, den, aux1_s, aux2_s
   double precision esc_1, esc_2, esc_3, esc_4, esc_5, num_escalar, den_escalar,
minor_d, minor_d2
   double precision  x(maxn,2), v(maxn,2), P_0(4,3), P_I(4,3), a(2), s(4),
points_distance(4) , X_0(4,2), n_g(4,2), pos_reflected(2), new_C0(2)
   integer,dimension (:,:),allocatable:: planes
   double precision,dimension(:,:),allocatable::  position_t0, position_t1, velocity_t1,
distance, xp_0
   character(len=3) stop, corner_reflection
   character(len=20) planecol, planecol2

   allocate(position_t0(maxn,2),  position_t1(maxn,2),  velocity_t1(maxn,2),
distance(maxn,5), planes(maxn,5), xp_0(maxn,2))

!----------------------Input files --------------------

   open(unit=43,file='../output/initial_position.dat')  ! particle positions at the beginning
of this iteration

   rewind(43)
   rewind(445)    ! vectors normal to the reservoir walls
   rewind(450)    ! points belonging to the reservoir walls

   do n_part=1,cont_domain
    read(43,*) (position_t0(n_part,j),j=1,2)
    xp_0(n_part,1) = position_t0(n_part,1)
    xp_0(n_part,2) = position_t0(n_part,2)
   end do
   close(43)

! Particle positions and velocities at the end of the iteration assuming that the particles did
not collide against the planes

   do n_part=1,cont_domain
    position_t1(n_part,1)= x(n_part,1)
    position_t1(n_part,2)= x(n_part,2)
    velocity_t1(n_part,1)= v(n_part,1)
    velocity_t1(n_part,2)= v(n_part,2)
   end do

!*******************************************************************
! Calculating the distance between the center of the particle in the initial position "C0"
! and in the final position "C1" and the planes
!*******************************************************************
   stop="NO"    !-------------------- Assumption that there were collisions in the first
numerical iteration

   do n_part=1,cont_domain       ! Loop for all particles
```

```
    cont=0
    distance(n_part,1) = n_part

!**********************************************************************
! Checking the possibilities of collisions for the particle "n_part"
!**********************************************************************
   do while (stop.eq."NO")   !AAA

! Calculations of the distance from the center "C1" of the particle "n_part" at the end
! point of its trajectory, after each possible collision, and each active plane:


!---- Distance to plane 01 (column 02 of the distance matrix):

    if (n_g(1,1).NE.0.or.n_g(1,2).NE.0) then
     aux_x=n_g(1,1)*(position_t1(n_part,1)-X_0(1,1))
     aux_y=n_g(1,2)*(position_t1(n_part,2)-X_0(1,2))
     distance(n_part,2)=aux_x+aux_y

    ! To ensure accuracy in collision detection

    distance(n_part,2) = ANINT(distance(n_part,2)*1000000.0)/1000000.0
   else
    distance(n_part,2)=99999.
   end if

!---- Distance to plane 02 (column 03 of the distance matrix):

   if (n_g(2,1).NE.0.or.n_g(2,2).NE.0) then
    aux_x=n_g(2,1)*(position_t1(n_part,1)-X_0(2,1))
    aux_y=n_g(2,2)*(position_t1(n_part,2)-X_0(2,2))

    distance(n_part,3)=aux_x+aux_y

   ! To ensure accuracy in collision detection:
    distance(n_part,3) = ANINT(distance(n_part,3)*1000000.0)/1000000.0
   else
    distance(n_part,3)=99999.
   end if

!--- Distance to plane 03 (column 04 of the distance matrix):

   if (n_g(3,1).ne.0.or.n_g(3,2).ne.0) then
    aux_x=n_g(3,1)*(position_t1(n_part,1)-X_0(3,1))
    aux_y=n_g(3,2)*(position_t1(n_part,2)-X_0(3,2))

    distance(n_part,4)=aux_x+aux_y

   ! To ensure accuracy in collision detection:
   distance(n_part,4) = ANINT(distance(n_part,4)*1000000.0)/1000000.0
```

```
    else
     distance(n_part,4)=99999.
    end if

!---- Distance to plane 04 (column 05 of the distance matrix):

    if (n_g(4,1).ne.0.or.n_g(4,2).ne.0) then
     aux_x=n_g(4,1)*(position_t1(n_part,1)-X_0(4,1))
     aux_y=n_g(4,2)*(position_t1(n_part,2)-X_0(4,2))
     distance(n_part,5)=aux_x+aux_y

     ! To ensure accuracy in collision detection:
     distance(n_part,5) = ANINT(distance(n_part,5)*1000000.0)/1000000.0
   else
     distance(n_part,5)=99999.
   end if

! ---- Initializing the "planes" matrix with zeros:

    planes(n_part,1)=n_part
    do j=2,5
     planes(n_part,j)=0  ! Assuming initially that there is no collision against any plane
    end do

!*****************************************************************************
! Performing the collision test, detecting and storing the planes against which collisions
! may have occurred
! Criterion: distances from the center of the particle at t0+dt and the plane considered
!*****************************************************************************
    do j=2,5

! If there is a possibility of collision against any plane, the position of the "planes" array
relative to the
!respective collision plane receives the value "one"

     if (distance(n_part,j).ge.0.and.distance(n_part,j).le.radius) then
      planes(n_part,j)=1
     else
       if (distance(n_part,j).LT.0) then
         planes(n_part,j)=1
       end if
      end if
    end do

!*****************************************************************************
! If there was a collision of the particle, it is necessary to check if it will still continue to
! collide
! with the planes within the planes within the time step
!*****************************************************************************
    stop="YES"    ! Assumpting there was no collision
```

```
  j=2
  do while (stop.eq."YES".and.j.le.5)
   if (planes(n_part,j).eq.1) then   ! If true, need to check for new collisions
    if (stop.eq."YES") then
     cont=cont+1          ! Record the possible number of collisions
     stop="NO"
    end if
   else
    j=j+1
   end if
  end do

!*********************************************************************
! Calculation of the "P_0" points of the "n_part" particle in relation to each possible
! collision plane
! Calculations performed only if there is a possibility of collision of the particles against
! one or more planes
!*********************************************************************
  if (stop.eq."NO") then       !!!!B

  open(unit=773,file='P_0.dat') ! file containing the coordinates of point P_0 (closest to
the collision plane)
  rewind(773)
  write (773,*) n_part
  do i=1,4      ! Initializing the matrix P_0 (referring to the particle "n_part")
   P_0(i,1)=i
    do j=2,3
     P_0(i,j)=0.
    end do      ! Columns 2 to 3 of the matrix are set to "zero"
  end do        ! It is assumed that there was no collision against any plane

!*********************************************************************
! Determining P_0 (position of the particle point that will touch each of the possible
! collision planes).
! Only if there is a possibility of collision
!*********************************************************************
  do i=2,5
   aux=planes(n_part,i)
   if (aux.ne.0) then
    if (i.eq.2) then      ! P_0 relative to plane 01 [Right side]
     P_0(i-1,2)=position_t0(n_part,1)+radius
     P_0(i-1,3)=position_t0(n_part,2)
    else
     if (i.eq.3) then    !P_0 relative to plane 02 [Left side]
      P_0(i-1,2)=position_t0(n_part,1)-radius
      P_0(i-1,3)=position_t0(n_part,2)
     else
      if (i.eq.4) then   !P_0 relative to plane 03  [Bottom]
       P_0(i-1,2)=position_t0(n_part,1)
       P_0(i-1,3)=position_t0(n_part,2)-radius
```

```
          else
           if (i.eq.5) then   !P_0 relative to plane 04  [Upper]
            P_0(i-1,2)=position_t0(n_part,1)
            P_0(i-1,3)=position_t0(n_part,2)+radius
           end if
         end if
        end if
      end if
    end if
  end do

  do i=1,4
   write(773,27) (P_0(i,j),j=1,3) ! Writing P_0 to an output file
27  format (3(2x,E13.6))
  end do
  close(773)

!************************************************************************
! Finding the directing vector "a" of the line cointaining the trajectory of "P_0"
! Calculate ! "a" from the positions of the centers of the particles at the beginning (C0)
! and at the end of the trajectory (C1)
! It is known  that the lines containing points P_0 and P_1 (line 1) and  the one containing
! points C0 and C1 are parallel
!************************************************************************
  dx=position_t1(n_part,1)-position_t0(n_part,1)
  dy=position_t1(n_part,2)-position_t0(n_part,2)

  module_a=sqrt(dx*dx+dy*dy)

! Obtaining the directing vector of the line that contains the trajectory of the particle until
the collision

  do i=1,2

   if ((module_a).ne.0) then    ! The particle moved
    a(i)=(position_t1(n_part,i)-position_t0(n_part,i))/module_a
   else                ! The particle is still
    a(i) = 0.
   end if

  end do

!************************************************************************
! Calculating the parameters of the line that intercepts the collision plane - equation of the
! line: P_I = P0 + s*a
!************************************************************************
  do i=1,4
   s(i)=99999.  ! If there is no collision against the plane i, the value of 's' relative to
the collision against that plane will be 99999.
  end do
```

```
do i=1,4
 if ((i.eq.1).and.(planes(n_part,i+1).ne.0)) then

 !***** Possible collision against the Plane 01 ***********************

 ! X_0 belongs to the plane and P_0 belongs to the line
 ! Calculating X_0-P_0:
   aux1_s=X_0(1,1)-P_0(i,2)
   aux2_s=X_0(1,2)-P_0(i,3)
 !  Taking the scalar product between (X_0-P_0) and the normal vector to the plane:
   esc_1=(aux1_s)*n_g(1,1)
   esc_2=(aux2_s)*n_g(1,2)
   num_escalar=esc_1+esc_2
 ! Making the scalar product between the direction vector of the line and the normal
vector to the plane:
   esc_4=a(1)*n_g(1,1)
  esc_5=a(2)*n_g(1,2)
  den_escalar=esc_4+esc_5

  if (den_escalar.ne.0) then
   s(1)=num_escalar/den_escalar
  else
   s(1) = 0.
  end if

 !---------- Collision point against the plane 01 --------------------
  P_I(i,1)=i
  P_I(i,2) = P_0(i,2) + s(1)*a(1)
  P_I(i,3) = P_0(i,3) + s(1)*a(2)

 else
  if ((i.eq.2).and.(planes(n_part,i+1).ne.0)) then

  !******Possible collision com o plane 02**********************

   aux1_s=X_0(2,1)-P_0(i,2)
   aux2_s=X_0(2,2)-P_0(i,3)

   esc_1=(aux1_s)*n_g(2,1)
   esc_2=(aux2_s)*n_g(2,2)

   num_escalar=esc_1+esc_2

   esc_4=a(1)*n_g(2,1)
   esc_5=a(2)*n_g(2,2)

   den_escalar=esc_4+esc_5

   if (den_escalar.ne.0) then
```

```
  s(2)=num_escalar/den_escalar
 else
  s(2) = 0.
 end if

 !---------- Collision point against the plane 02---------------------
 P_I(i,1)=i
 P_I(i,2) = P_0(i,2) + s(2)*a(1)
 P_I(i,3) = P_0(i,3) + s(2)*a(2)

 else
  if ((i.eq.3).and.(planes(n_part,i+1).ne.0)) then

  !**********Possible collision against the Plane 03*******************

   aux1_s=X_0(3,1)-P_0(i,2)
   aux2_s=X_0(3,2)-P_0(i,3)
   esc_1=(aux1_s)*n_g(3,1)
   esc_2=(aux2_s)*n_g(3,2)

   num_escalar=esc_1+esc_2
   esc_4=a(1)*n_g(3,1)
   esc_5=a(2)*n_g(3,2)

   den_escalar=esc_4+esc_5

  if (den_escalar.ne.0) then
   s(3)=num_escalar/den_escalar
  else
   s(3) = 0.
  end if

 !---------- Collision point against the plane 03---------------------
  P_I(i,1)=i
  P_I(i,2) = P_0(i,2) + s(3)*a(1)
  P_I(i,3) = P_0(i,3) + s(3)*a(2)

  else
   if ((i.eq.4).and.(planes(n_part,i+1).ne.0)) then

  !******** Possible collision against the Plane 04**************

    aux1_s=X_0(4,1)-P_0(i,2)
    aux2_s=X_0(4,2)-P_0(i,3)

    esc_1=(aux1_s)*n_g(4,1)
    esc_2=(aux2_s)*n_g(4,2)
    num_escalar=esc_1+esc_2

    esc_4=a(1)*n_g(4,1)
```

```
            esc_5=a(2)*n_g(4,2)

            den_escalar=esc_4+esc_5

           if  (den_escalar.ne.0) then
             s(4)=num_escalar/den_escalar
           else
             s(4) = 0.
           end if

          !---------- Collision point against the plane 04 -----------------
          P_I(i,1)=i
          P_I(i,2) = P_0(i,2) + s(4)*a(1)
          P_I(i,3) = P_0(i,3) + s(4)*a(2)

          end if
        end if
      end if
     end if
    end do
!********************************************************************
! Calculating (P_I-P_0), in absolute value, for possible collision planes
! (Checking between all possible collision planes)
!********************************************************************
    do i=1,4
     points_distance(i)=abs(s(i))   ! The distance between points P_0 and P_I is equal to
the absolute value of s(i)
    end do

!********************************************************************
! Checking the shortest distance between points on the trajectory to the collision and
! determining the collision plane(s)!
!********************************************************************
   stop="YES"     ! Assumpting there was no collision

   minor_d=99999.
   minor_d2=99999.
   ncol=0
   plane_col=0     ! Assuming there was no collision against any plane
   plane_col2=0   ! There may be simultaneous collision between two planes (corner)

   do i=1,4    ! In this loop, the collision planes are identified
    if (points_distance(i).lt.minor_d) then
     ncol=1
     plane_col=i  !If there was a collision, storing the collision plane
     minor_d= points_distance(i)
    else
     if ((ncol.eq.1).and.(points_distance(i).eq.minor_d)) then  ! If there is a collision at
the corner (meeting between two planes)
      ncol=ncol+1
```

```
      plane_col2=i
      minor_d2=points_distance(i)
     end if
   end if
 end do

 if (ncol.ne.0) then
   if ((minor_d.lt.minor_d2)) then
    ncol=1
   else
    if ((minor_d2.lt.minor_d)) then
     ncol=1
     plane_col=plane_col2
     minor_d=minor_d2
    else    ! Collision of the particle with two planes simultaneously
     if ((minor_d.eq.minor_d2)) then
      ncol=2
     end if
    end if
   end if
 end if

!**********************************************************************
! Just checking the collision of the particle "n_part"
!**********************************************************************
  if (ncol.ne.0) then  !!!E

   if (ncol.eq.1) then  !!!F   ! If there was a collision against a plane, the particle position
and velocity must be updated.
    stop="NO"

    if (plane_col.eq.1) then
      planecol="[1] - [RIGHT PLANE]"
     else
      if (plane_col.eq.2) then
        planecol="[2] - [LEFT PLANE]"
      else
        if (plane_col.eq.3) then
          planecol="[3] - [BOTTOM]"
        else
          if (plane_col.eq.4) then
            planecol= "[4] - [UPPER PLANE]"
          end if
        end if
      end if
    end if

    write(66,*) '-----------------------------------------------------------------'
    write(66,*) 'ITERATION = ' , itimestep
    write(66,*) 'Particle', n_part,'  had contact with the plane ', planecol
```

```
    write(66,*) 'Initial position = ', position_t0(n_part,1), position_t0(n_part,2)
    write(66,*) ''

!***********************************************************************
! PERFORMING THE REFLECTION OF THE PARTICLE "n_part"
! Reflecting the final position of the center to the interior of the domain
!***********************************************************************
   if (plane_col.eq.1) then  ! collision against the plane 01

    if (CR.gt.0.) then

      pos_reflected(1)=position_t1(n_part,1)+((1.      +      CR      )*(radius-
distance(n_part,2)))*n_g(1,1)
      pos_reflected(2)=position_t1(n_part,2)+(2*(radius-distance(n_part,2)))*n_g(1,2)

      velocity_t1(n_part,1)=(-1)*CR*velocity_t1(n_part,1) !x-direction
    else
      pos_reflected(1)=P_I(1,2)-radius
      pos_reflected(2)=P_I(1,3)

      velocity_t1(n_part,1)=(-1.)*CR*velocity_t1(n_part,1)  !x-direction

    end if

!***********************************************************************
! Correcting particle velocity and updating velocity matrix after each !collision
!***********************************************************************
   else
    if (plane_col.eq.2) then  ! collision against the plane 02

      if (CR.gt.0.) then
        pos_reflected(1)=position_t1(n_part,1)+((1.      +      CR      )*(radius-
distance(n_part,3)))*n_g(2,1)
        pos_reflected(2)=position_t1(n_part,2)+(2*(radius-distance(n_part,3)))*n_g(2,2)

       ! Updating the velocity at the end of the iteration
        velocity_t1(n_part,1)=(-1)*CR*velocity_t1(n_part,1)   !x-direction

      else
       pos_reflected(1)=P_I(2,2)+radius   ! position of the center of mass
       pos_reflected(2)=P_I(2,3)

       velocity_t1(n_part,1)=(-1.)*CR*velocity_t1(n_part,1)   !x-direction

      end if

    else
     if (plane_col.eq.3) then  ! collision against the plane 03

      if (CR.gt.0.) then
```

```
            pos_reflected(1)=position_t1(n_part,1)+(2*(radius-
distance(n_part,4)))*n_g(3,1)
            pos_reflected(2)=position_t1(n_part,2)+((1.        +        CR        )*(radius-
distance(n_part,4)))*n_g(3,2)

            velocity_t1(n_part,2)=(-1)*CR*velocity_t1(n_part,2)  !y-direction

          else
            pos_reflected(1)=P_I(3,2)
            pos_reflected(2)=P_I(3,3)+radius

            velocity_t1(n_part,2)=(-1.)*CR*velocity_t1(n_part,2)  !y-direction
          end if

        else
          if (plane_col.eq.4) then  ! collision against the plane 04

            if (CR.GT.0.) then
              pos_reflected(1)=position_t1(n_part,1)+(2*(radius-
distance(n_part,5)))*n_g(4,1)
              pos_reflected(2)=position_t1(n_part,2)+((1.        +        CR        )*(radius-
distance(n_part,5)))*n_g(4,2)

              velocity_t1(n_part,2)=(-1)*CR*velocity_t1(n_part,2)  !y-direction

            else
              pos_reflected(1)=P_I(4,2)
              pos_reflected(2)=P_I(4,3)-radius

              velocity_t1(n_part,2)=(-1.)*CR*velocity_t1(n_part,2)  !y-direction

            end if

          end if
        end if
      end if
    end if

!*****************************************************************
! Obtaining the new position of the center of the particle "new C_0" for the new trajectory
! after the previous collisions
! This is the position of the mass center of the particle at the instant of the previous
! collision against the plane
!*****************************************************************

    new_C0(1)=position_t0(n_part,1)+s(plane_col)*a(1)
    new_C0(2)=position_t0(n_part,2)+s(plane_col)*a(2)
!*****************************************************************
! Updating the initial position matrix of the particle "n_part". This is the position of the
particle center for the next iteration
```

```
!*********************************************************************
    do j=1,2
     position_t0(n_part,j)=new_C0(j)
    end do

!*********************************************************************
! Updating the matrix with the final position of the particle "n_part" from the coordinates
! after reflection
! This is the final position of the particle mass centre after the collision. If there are no
!  more collisions,its position ! is already !defined.
!*********************************************************************
    do j=1,2
      position_t1(n_part,j)=pos_reflected(j)
    end do

!*********************************************************************
!   COLLISION AGAINST TWO PLANES SIMULTANEOUSLY
!*********************************************************************
   else   !!!F
    if (ncol.eq.2) then
      stop="NO"
      corner_reflection="NO"

      if (plane_col.eq.1) then
        planecol="[1] - [RIGHT PLANE]"
      else
        if (plane_col.eq.2) then
        planecol="[2] - [LEFT PLANE]"
        else
          if (plane_col.eq.3) then
            planecol="[3] - [BOTTOM] "
          else
            if (plane_col.eq.4) then
              planecol=" [4] - [UPPER PLANE] "
            end if
          end if
        end if
      end if

      if (plane_col2.eq.1) then
        planecol2="[1] - [RIGHT PLANE]"
      else
        if (plane_col2.eq.2) then
          planecol2="[2] - [LEFT PLANE]"
        else
          if (plane_col2.eq.3) then
            planecol2="[3] - [BOTTOM] "
          else
            if (plane_col2.eq.4) then
              planecol2=" [4] - [UPPER PLANE] "
```

```
        end if
       end if
      end if
     end if

     write(66,*) '--------------------------------------------------------------------------'
     write(66,*) 'ITERATION = ' , itimestep
     write(66,*) 'Particle ', n_part,' had contact with the planes ', planecol, ' and ',
planecol2
     write(66,*) 'Initial position = ', position_t0(n_part,1), position_t0(n_part,2)
     write(66,*) ' '

!*****************************************************************************
! Performing the reflection of the particle "n_part"
! Reflecting the final position of the center to the interior of the !domain
!*****************************************************************************

!-------------- Collision against the first plane ---------------------------------------------------

     if (plane_col.eq.1) then   ! Collision against the plane 01

      if (CR.gt.0.) then
      ! Reflecting the final position of the center of mass into the interior of the domain
        pos_reflected(1)=position_t1(n_part,1)+((1.      +       CR       )*(radius-
distance(n_part,2)))*n_g(1,1)
        pos_reflected(2)=position_t1(n_part,2)+(2*(radius-distance(n_part,2)))*n_g(1,2)
      else
        pos_reflected(1)=P_I(1,2)-radius
        pos_reflected(2)=P_I(1,3)
      end if

!*****************************************************************************
! Correcting particle velocity and updating velocity matrix after each !collision
!*****************************************************************************
      velocity_t1(n_part,1)=(-1.)*CR*velocity_t1(n_part,1)

     else
      if (plane_col.eq.2) then ! Collision against the plane 02

       if (CR.gt.0.) then
        pos_reflected(1)=position_t1(n_part,1)+((1.       +        CR        )*(radius-
distance(n_part,3)))*n_g(2,1)
        pos_reflected(2)=position_t1(n_part,2)+(2*(radius-distance(n_part,3)))*n_g(2,2)
       else
        pos_reflected(1)=P_I(2,2)+radius
        pos_reflected(2)=P_I(2,3)
       end if
      ! Updating the velocity at the end of the iteration
        velocity_t1(n_part,1)=(-1.)*CR*velocity_t1(n_part,1)
```

```
    else
     if (plane_col.eq.3) then  ! Collision against the plane 03

      if (CR.gt.0.) then
        pos_reflected(1)=position_t1(n_part,1)+(2*(radius-
distance(n_part,4)))*n_g(3,1)
         pos_reflected(2)=position_t1(n_part,2)+((1.     +     CR     )*(radius-
distance(n_part,4)))*n_g(3,2)
      else
        pos_reflected(1)=P_I(3,2)
        pos_reflected(2)=P_I(3,3)+radius
     end if

     ! Updating the velocity at the end of the iteration
     velocity_t1(n_part,2)=(-1)*CR*velocity_t1(n_part,2)

      else
       if (plane_col.eq.4) then  ! Collision against the plane 04

        if (CR.gt.0.) then
              pos_reflected(1)=position_t1(n_part,1)+(2*(radius-
distance(n_part,5)))*n_g(4,1)
              pos_reflected(2)=position_t1(n_part,2)+((1.    +    CR    )*(radius-
distance(n_part,5)))*n_g(4,2)
        else
          pos_reflected(1)=P_I(4,2)
          pos_reflected(2)=P_I(4,3)-radius
        end if

        !  Updating the velocity at the end of the iteration
        velocity_t1(n_part,2)=(-1)*CR*velocity_t1(n_part,2)

       end if
     end if
    end if
   end if

!***********************************************************************
! Updating the final position matrix of the particle "n_part" from the coordinates after the
! first reflection
!***********************************************************************
   do j=1,2
    position_t1(n_part,j)=pos_reflected(j)
   end do

!--------------- Collision against the second plane -------------------------------------------

  if (plane_col2.eq.1) then  ! Collision against the plane 01

   if (CR.gt.0.) then
```

```
      pos_reflected(1)=position_t1(n_part,1)+((1.          +          CR          )*(radius-
distance(n_part,2)))*n_g(1,1)
      pos_reflected(2)=position_t1(n_part,2)+(2*(radius-distance(n_part,2)))*n_g(1,2)
     else
      pos_reflected(1)=P_I(1,2)-radius
      pos_reflected(2)=P_I(1,3)
     end if
!**********************************************************************
! Correcting particle velocity and updating velocity matrix after each collision
!**********************************************************************
     velocity_t1(n_part,1)=(-1.)*CR*velocity_t1(n_part,1)

    else
     if (plane_col2.eq.2) then     ! Collision against the plane 02

      if (CR.gt.0.) then
       pos_reflected(1)=position_t1(n_part,1)+((1.          +          CR          )*(radius-
distance(n_part,3)))*n_g(2,1)
       pos_reflected(2)=position_t1(n_part,2)+(2*(radius-
distance(n_part,3)))*n_g(2,2)
      else
       pos_reflected(1)=P_I(2,2)+radius
       pos_reflected(2)=P_I(2,3)
      end if

!**********************************************************************
!       Updating velocity matrix
!**********************************************************************
     velocity_t1(n_part,1)=(-1.)*CR*velocity_t1(n_part,1)

    else
     if (plane_col2.eq.3) then     ! Collision against the plane 03

      if (CR.gt.0.) then
       pos_reflected(1)=position_t1(n_part,1)+(2*(radius-distance(n_part,4)))*n_g(3,1)
       pos_reflected(2)=position_t1(n_part,2)+((1.          +          CR          )*(radius-
distance(n_part,4)))*n_g(3,2)

      else
       pos_reflected(1)=P_I(3,2)
       pos_reflected(2)=P_I(3,3)+radius
      end if

!**********************************************************************
!       Updating velocity matrix
!**********************************************************************
     velocity_t1(n_part,2)=(-1.)*CR*velocity_t1(n_part,2)
    else
     if (plane_col2.eq.4) then  ! Collision against the plane 04
```

```
        if (CR.gt.0.) then
         pos_reflected(1)=position_t1(n_part,1)+(2*(radius-
distance(n_part,5)))*n_g(4,1)
         pos_reflected(2)=position_t1(n_part,2)+((1.      +      CR      )*(radius-
distance(n_part,5)))*n_g(4,2)

        else
         pos_reflected(1)=P_I(4,2)
         pos_reflected(2)=P_I(4,3)-radius
        end if

!*******************************************************************
!     Updating velocity matrix
!*******************************************************************
        velocity_t1(n_part,2)=(-1.)*CR*velocity_t1(n_part,2)

       end if
      end if
     end if
 end if

!*******************************************************************
! The next commands are used to remove the infinite loop if the reflection
! of the particle puts it in contact with a corner (encounter between 2 planes)
*******************************************************************
 if ( (distance(n_part,2).eq.radius).and.(distance(n_part,4).eq.radius)) then
      corner_reflection = "YES"
    else
       if ( (distance(n_part,2).eq.radius).and.(distance(n_part,5).eq.radius)) then
        corner_reflection = "YES"
        else
         if ( (distance(n_part,3).eq.radius).and.(distance(n_part,4).eq.radius)) then
        corner_reflection = "YES"
         else
          if ( (distance(n_part,3).eq.radius).and.(distance(n_part,5).eq.radius)) then
            corner_reflection = "YES"
          end if
         end if
       end if
     end if

!-------------------------------------------------------------------
!  Finding the position of the particle center at the moment of collision
!-------------------------------------------------------------------
     do i=1,2
      new_C0(i)=position_t0(n_part,i)
     end do
     if ((plane_col.eq.1).or.(plane_col.eq.2)) then
      new_C0(1)=position_t0(n_part,1)+s(plane_col)*a(1)
     else
```

```
    if ((plane_col.eq.3).or.(plane_col.eq.4)) then
     new_C0(2)=position_t0(n_part,2)+s(plane_col)*a(2)
    end if
   end if

   if ((plane_col2.eq.1).or.(plane_col2.eq.2)) then
    new_C0(1)=position_t0(n_part,1)+s(plane_col2)*a(1)
   else
    if ((plane_col2.eq.3).or.(plane_col2.eq.4)) then
     new_C0(2)=position_t0(n_part,2)+s(plane_col2)*a(2)
    end if
   end if

!**********************************************************************
! Updating the initial position matrix of the particle "n_part". This is the position of the
! particle centre for the next iteration
!**********************************************************************
   do j=1,2
    position_t0(n_part,j)=new_C0(j)
   end do

!**********************************************************************
! Updating the final position matrix of the particle "n_part" from the coordinates after the
! second reflection
!**********************************************************************
   do j=1,2
    position_t1(n_part,j)=pos_reflected(j)
   end do

  end if  ! collision against 02 planes
end if    ! collision against only 01 plane     !!!F

end if !!!E

else !!!!B   ! It will only enter this ELSE if there is no more collision (distances to planes
greater than zero)

 stop="YES"

end if !!!!! B

 if (corner_reflection.eq."YES") then
    stop = "YES"
 end if

 ! Stopping criterion: if the particle had contacted with the wall and is at rest
 if ( (velocity_t1(n_part,1).EQ.0.).and.(velocity_t1(n_part,2).EQ.0.) ) then
    stop = "YES"
 end if
```

```
end do ! AAA do while

stop="NO"
!-----------------------------------------------------------------------------------------------------
! Great! All collisions for the particle "n_part"  have been performed at the current time
! step.  Now, let's change the particle
!-----------------------------------------------------------------------------------------------------

end do  ! Particle change

write(67,*) '-----------------------ITERATION', itimestep, '-------------------------'
do n_part=1,cont_domain
  write(67,*) '----------------------------------------------------------------------'
  write(67,*) 'Final reflected position of center of mass (Cf):'
  write(67,*) 'Particle = ', n_part
  write(67,*) 'x = ' , position_t1(n_part,1), ' - y = ' , position_t1(n_part,2)
  write(67,*) ' '
  write(67,*) '----------------------------------------------------------------------'
end do

do n_part=1,cont_domain
  write(68,*) '-------------------------------------------------------------'
  write(68,*) n_part, (position_t1(n_part,1) - xp_0(n_part,1)), (position_t1(n_part,2) -
xp_0(n_part,2))
  write(68,*) ' '
end do

!-----------------------------------------------------------------------
! ----- Checking if static equilibrium has been maintained
!-----------------------------------------------------------------------

do i=1,cont_domain

  if ( ( (position_t1(i,1) - xp_0(i,1) ).EQ.0.0) .AND. ((position_t1(i,2) - xp_0(i,2)).EQ.0.0
)) then
    write(69,*)  ' '
    change = 0
    write(69,*)   'THERE WAS NO CHANGE IN THE POSITIONS OF THE
PARTICLES IN THE ITERATION = ', itimestep
    write(69,*)  ' '
  else
    change = 1
    write(69,*)  'ITERATION = ', itimestep
    write(69,*)  'CHANGE IN PARTICLE POSITION – PARTICLE ', i
    write(69,*)  'delta_x = ',  (position_t1(i,1) - xp_0(i,1))
    write(69,*)  'delta_y = ',  (position_t1(i,2) - xp_0(i,2))
    write(69,*)  ' '
  end if
end do
```

```
write(455,*) '-----------------------------------------------------------------'
write(455,*) '      FINAL RESULTS FOR THE ITERATION        '
write(455,*) '-----------------------------------------------------------------'
do i=1,cont_domain
 write(455,*) 'Final reflected position of the center of mass (Cf):'
 write(455,*) i, (position_t1(i,j),j=1,2)
 write(455,*) ' '
end do

do i=1,cont_domain
 write(455,*) 'Final velocity of the centre of mass (Vf):'
 write(455,*) i, (velocity_t1(i,j),j=1,2)
 write(455,*) ' '
end do

write(455,*) '-------------------------------------------------'
write(455,*) ''

! Updating positions and velocities for the next iteration:

do i=1,cont_domain
 x(i,1) = position_t1(i,1)
 x(i,2) = position_t1(i,2)
 v(i,1) = velocity_t1(i,1)
 v(i,2) = velocity_t1(i,2)
end do

end subroutine collision_test


 subroutine output(cont_domain, x, v, accel, p, itimestep)

!-----------------------------------------------------------------------------------------
! Subroutine for saving particles physical properties to files
!
! x -- coordinates of particles                                              [in]
! v -- velocities of particles                                               [in]
! cont_domain -- total particle number                                       [in]
!-----------------------------------------------------------------------------------------

   implicit none
   include 'param.inc'

   integer aux_p, d, aux, n_part, k, funit, funit2, funit3, funit4, itimestep, cont_domain,
record
   double precision x(maxn,2), v(maxn,2), accel(maxn,2), p(maxn)
   character(len=25)::aux2
   character(len=45)::name_pos, name_vel, name_accel, name_pressure
```

```
    record = 100   ! step to record output files

    aux=itimestep

    if ((itimestep.eq.1).or.(mod(itimestep,record).eq.0)) then

      write(aux2,*) itimestep   ! Converting an integer value to a character

        if (itimestep.eq.1) then
          funit=itimestep+3100
          funit2=itimestep+6100
          funit3=itimestep+10100
          funit4=itimestep+15100
        else
          funit=int(itimestep/100)+3100
          funit2=int(itimestep/100)+6100
          funit3=int(itimestep/100)+10100
          funit4=itimestep+15100
        end if


        if (aux.lt.10) name_pos='../output/data/position/000000'//trim(adjustl(aux2))//'.dat'
        if                                          ((aux.ge.10).and.(aux.lt.100))
name_pos='../output/data/position/00000'//trim(adjustl(aux2))//'.dat'
        if                                          ((aux.ge.100).and.(aux.lt.1000))
name_pos='../output/data/position/0000'//trim(adjustl(aux2))//'.dat'
        if                                          ((aux.ge.1000).and.(aux.lt.10000))
name_pos='../output/data/position/000'//trim(adjustl(aux2))//'.dat'
        if                                          ((aux.ge.10000).and.(aux.lt.100000))
name_pos='../output/data/position/00'//trim(adjustl(aux2))//'.dat'
        if                                          ((aux.ge.100000).and.(aux.lt.1000000))
name_pos='../output/data/position/0'//trim(adjustl(aux2))//'.dat'
        if (aux.eq.1000000) name_pos='../output/data/position'//trim(adjustl(aux2))//'.dat'

        open(unit=funit,file=name_pos)
        do n_part = 1, cont_domain
          write (funit,221) n_part, x(n_part,1), x(n_part,2)
        end do

!---------------------------------------------------------------------

        if (aux.lt.10)  name_vel='../output/data/velocity/000000'//trim(adjustl(aux2))//'.dat'
        if                                          ((aux.ge.10).and.(aux.lt.100))
name_vel='../output/data/velocity/00000'//trim(adjustl(aux2))//'.dat'
        if                                          ((aux.ge.100).and.(aux.lt.1000))
name_vel='../output/data/velocity/0000'//trim(adjustl(aux2))//'.dat'
        if                                          ((aux.ge.1000).and.(aux.lt.10000))
name_vel='../output/data/velocity/000'//trim(adjustl(aux2))//'.dat'
        if                                          ((aux.ge.10000).and.(aux.lt.100000))
name_vel='../output/data/velocity/00'//trim(adjustl(aux2))//'.dat'
```

```
 if ((aux.ge.100000).and.(aux.lt.1000000))
name_vel='../output/data/velocity/0'//trim(adjustl(aux2))//'.dat'
 if (aux.eq.1000000) name_vel='../output/data/velocity'//trim(adjustl(aux2))//'.dat'

 open(unit=funit2,file=name_vel)

 do n_part = 1, cont_domain
  write (funit2,221) n_part, v(n_part,1), v(n_part,2)
 end do

!-----------------------------------------------------------------------

 if (aux.lt.10)
name_accel='../output/data/acceleration/000000'//trim(adjustl(aux2))//'.dat'
 if ((aux.ge.10).and.(aux.lt.100))
name_accel='../output/data/acceleration/00000'//trim(adjustl(aux2))//'.dat'
 if ((aux.ge.100).and.(aux.lt.1000))
name_accel='../output/data/acceleration/0000'//trim(adjustl(aux2))//'.dat'
 if ((aux.ge.1000).and.(aux.lt.10000))
name_accel='../output/data/acceleration/000'//trim(adjustl(aux2))//'.dat'
 if ((aux.ge.10000).and.(aux.lt.100000))
name_accel='../output/data/acceleration/00'//trim(adjustl(aux2))//'.dat'
 if ((aux.ge.100000).and.(aux.lt.1000000))
name_accel='../output/data/acceleration/0'//trim(adjustl(aux2))//'.dat'
 if (aux.eq.1000000)
name_accel='../output/data/acceleration'//trim(adjustl(aux2))//'.dat'

 open(unit=funit3,file=name_accel)

 do n_part = 1, cont_domain
  write (funit3,221) n_part, accel(n_part,1), accel(n_part,2)
 end do

221 format (I8, 2(1x,D21.14))

!-----------------------------------------------------------------------

 if (aux.lt.10)
name_pressure='../output/data/pressure/000000'//trim(adjustl(aux2))//'.dat'
 if ((aux.ge.10).and.(aux.lt.100))
name_pressure='../output/data/pressure/00000'//trim(adjustl(aux2))//'.dat'
 if ((aux.ge.100).and.(aux.lt.1000))
name_pressure='../output/data/pressure/0000'//trim(adjustl(aux2))//'.dat'
 if ((aux.ge.1000).and.(aux.lt.10000))
name_pressure='../output/data/pressure/000'//trim(adjustl(aux2))//'.dat'
 if ((aux.ge.10000).and.(aux.lt.100000))
name_pressure='../output/data/pressure/00'//trim(adjustl(aux2))//'.dat'
 if ((aux.ge.100000).and.(aux.lt.1000000))
name_pressure='../output/data/pressure/0'//trim(adjustl(aux2))//'.dat'
```

```
       if                                                       (aux.eq.1000000)
name_pressure='../output/data/pressure'//trim(adjustl(aux2))//'.dat'

       open(unit=funit4,file=name_pressure)

        do n_part = 1, cont_domain
          write (funit4,222) n_part, p(n_part)
        end do

 222   format (I8, 1x, D21.14)

       close (funit)
       close (funit2)
       close(funit3)
       close(funit4)

   end if

end subroutine output



!------------------------------------------------------------------------------------------------
! Here are the parameters and constants used in the SPH software
! A file must be created with the name param.inc
!------------------------------------------------------------------------------------------------

!   maxn    : Maximum number of particles
!   max_interation : Maximum number of interaction pairs
     integer maxn,max_interaction, n_neighbours
     parameter ( maxn = 100000, n_neighbours = 50 )

!   SPH algorithm for calculete the pressure force
!   pa_sph = 1 : (1/rho(i)/ (p(j) -p(i) )
!          2 : ((p(i)/rho(i)**2+p(j)/rho(j)**2)
     integer pa_sph
     parameter(pa_sph = 2)

     double precision pi
     parameter ( pi = 3.14159265358979323846 )

!   Smoothing kernel function
!   skf = 1, Cubic spline kernel (Monaghan 1985)
!       = 2, Quintic spline kernel (Morris 1997)
!       = 3,  Quartic kernel (Liu & Lam)
     integer skf
     parameter(skf = 1)

# ==========================================================
```

```
# makefile
# ===================================================================
/folder_address_of_item_2/fonts/main.run:   main.o  input.o  geometry.o  direct_find.o
modified_pressure.o absolute_pressure.o  \
kernel.o   internal_force.o   external_force.o   acceleration.o   time_integration.o
collision_test.o output.o \

#
	ifort -O3 -o / folder address of item 1/fonts/main.run  main.o input.o geometry.o
direct_find.o modified_pressure.o  \
absolute_pressure.o   kernel.o   internal_force.o   external_force.o   acceleration.o
time_integration.o collision_test.o output.o \

#
#

main.o:          param.inc  main.f90
	ifort -O3 -c  main.f90
input.o:  param.inc input.f90
	ifort -O3 -c  input.f90
geometry.o:
	ifort -O3 -c  geometry.f90
direct_find.o:
	ifort -O3 -c  direct_find.f90
modified_pressure.o:
	ifort -O3 -c  modified_pressure.f90
absolute_pressure.o:
	ifort -O3 -c  absolute_pressure.f90
kernel.o:    param.inc kernel.f90
	ifort -O3 -c  kernel.f90
internal_force.o:   param.inc internal_force.f90
	ifort -O3 -c  internal_force.f90
external_force.o:
	ifort -O3 -c  external_force.f90
acceleration.o:   param.inc acceleration.f90
	ifort -O3 -c  acceleration.f90
time_integration.o:  param.inc time_integration.f90
	ifort -O3 -c  time_integration.f90
collision_test.o:
	ifort -O3 -c  collision_test.f90
output.o:
	ifort -O3 -c  output.f90
# ===================================================================
#  clean
# ===================================================================
clean:
	rm -rf *.o
```

Guidelines for implementing the program:

1. Create individual files (*name.f90*) for the *main program* and *subroutines*, besides *param.inc* and *Makefile*
2. Create a folder with the chosen name to paste all files
3. Inside the previous folder, create the *fonts*, *output* and *figures* folders
4. Inside *output*, create the *data* folder with the folders named *acceleration*, *position*, *pressure and velocity*
5. Create a file named *Makefile* with the content presented above
6. Set the folder address of item 2 in the *Makefile*
7. Open the terminal inside */folder_address_of item_2/fonts/* and give the command make
8. Run the program with the command ./main.run

Note: Inside the output folder there is a script to generate graphs showing the evolution of fluid particles over time.

Chapter 3
Two-Dimensional Dam-Breaking over a Dry Bed

```
program main

!  A file named "main.f90" can be created

!-----------------------------------------------------------------------------------------------
!  mass -- mass of particles                                              [in]
!  ntotal -- total particle number                                        [in]
!  dt -- Time step                                                        [in]
!  x -- coordinates of particles                                      [in/out]
!  rho -- densities of particles                                      [in/out]
!  p -- pressure  of particles                                        [in/out]
!  hsml -- smoothing lengths of particles                             [in/out]
!-----------------------------------------------------------------------------------------------

    implicit none
    include 'param.inc'

    integer itimestep, maxtimestep, ntotal, d, m,  i, k,  hours, minutes, seconds,
hours_in_seconds, nx, ny, mp, np, recording_step, aux
    double precision dt, factor, start, finish, time, radius, xl, yl,xlmin, xlmax, ylmin,
ylmax, dx1, kh, K_scale, vmax, grav, Mach, b,  c_sound, CR
    integer, dimension (:,:), allocatable:: neighbours
    integer, dimension (:), allocatable:: surface_particles
    double precision, dimension (:,:), allocatable:: x, v,gravity,r, dx, dy, w, dwdx, dwdy,
dwdr, viscous_force, pressure_force, accel, v_xsph,  n_g, X_0
    double precision, dimension (:), allocatable ::  mass, rho, p, hsml, viscosity,
sum_CSPM,  drhodt

   open(445,file='../output/normals.dat')      ! vectors normal to the reservoir wall
   open(450,file='../output/flat_points.dat')    ! points belonging to the reservoir walls
   open(455,file='../output/positions_and_velocities_of_centres_of_mass.dat')
   open(998, file='../output/geometry.dat')
   open(999, file='../output/initial_distribution.dat')
   open(1000, file='../output/timestep.dat')

!-----------------------------------------------------------------------------------------------
!------------------------------------------ INPUT DATA -----------------------------------------
!-----------------------------------------------------------------------------------------------

 !  GEOMETRY TO BE SIMULATED:
 !  --- Tank dimensions:
   xlmin=0.
   ylmin=0.
```

Supplementary Information The online version contains supplementary material available at https://doi.org/10.1007/978-3-031-99370-1_3.

C. A. Dutra Fraga Filho, *Computational Algorithms for Reflective Boundary Conditions in SPH Fluid Dynamics Simulation*, Synthesis Lectures on Mechanical Engineering, https://doi.org/10.1007/978-3-031-99370-1_3

```
    xlmax=0.42
    ylmax=0.44
    grav=9.81      ! gravity magnitude

    write(998,554) xlmin, xlmax
    write(998,554) ylmin, ylmax

554 format(1x,4(E12.4))
   close(998)

 ! --- Fluid inside the tank:
    xl=0.114
    yl=2*xl

    m=5              ! number of particles per side of the unit cell containing water
    nx=7             ! number of cells in the x-direction
    ny = nx*int(yl/xl)  ! number of cells in the x-direction
    mp = m*nx+1        ! number of particles per domain line
    np = m*ny+1        ! number of particles per domain column
    ntotal = mp*np     ! total number of particles

    vmax= sqrt(2*grav*yl)     ! maximum fluid velocity
    Mach= 0.10                ! Mach number
    c_sound=vmax/Mach         ! sound velocity in the fluid

    dt = 1.e-5        ! time step (s)
    CR = 1.00          ! Coefficient of restitution of kinetic energy (reflective boundary
conditions)
    recording_step = 100  ! The step for recording the output files

!---------------------------------------------------------------------------------------

    allocate (neighbours(maxn,n_neighbours), surface_particles(maxn), x(maxn,2),
v(maxn,2), gravity(maxn,2), r(maxn,n_neighbours), dx(maxn, n_neighbours), dy(maxn,
n_neighbours), w(maxn, n_neighbours), dwdx(maxn, n_neighbours), dwdy(maxn,
n_neighbours), dwdr(maxn, n_neighbours), viscous_force(maxn,2),
pressure_force(maxn,2), accel(maxn,2), v_xsph(maxn,2), n_g(4,2), X_0(4,2),
mass(maxn),rho(maxn), p(maxn), hsml(maxn),
viscosity(maxn),sum_CSPM(maxn),drhodt(maxn))

    print*, ''
    print*, '--------------------------------------------------------------------------------'
    print*, '          WELCOME TO THE START OF THE SIMULATION                                '
    print*, '--------------------------------------------------------------------------------'
    print*, ''
    print*, ' Tank lenght (m) = ', xlmax
    print*, ''
    print*, ' Width of water in tank (m) = ', xl
    print*, ''
    print*, ' Water column height (m) = ' , yl
```

```
    print*, ' '
    print*, ' Total number of particles in the domain = '
    print*, mp, 'x', np , ' = ', np*mp
    print*, ' '
    print*, '-------------------------------------------------------------------------------------------'
    print*, ' '

    call cpu_time(start)

    rewind(998)
    rewind(999)
    rewind(1000)
    write(999,*) ntotal
    write(1000,*) dt
    write(1000,*) recording_step
    write(1000,*) maxtimestep
    close(999)
    close(1000)

    call geometry (xlmax, ylmax, n_g, X_0)

    call input(ntotal,xl, yl, x,m, mp, np, nx, dx1, kh, mass, rho, hsml, p, viscosity, K_scale,
surface_particles)

!------------------------ CREATING FILE FOR POST-PROCESSING ----------------------
     open(17,file='../output/matlab_parameters.dat')
     rewind(17)
     write(17,18) xlmin, xlmax, ylmin, ylmax
     write(17,19) ntotal, 0, 0, 0
18   format (4(1x,e10.3))
19   format (4(1x,I8))
     close(17)
!-----------------------------------------------------------------------------------------------------
    radius = dx1/2.  ! particle radius
    aux=1
    itimestep = 1
    maxtimestep = 30000    ! Simulated physical time of 0.30 s

    do i=aux,maxtimestep   ! Simulation until the end of the defined physical time

      if (itimestep.ne.1) call h_upgrade(dt, ntotal, rho, drhodt, hsml)

      call direct_find(itimestep, ntotal, hsml, neighbours, x, r, dx, dy, K_scale)

      call kernel(ntotal, r, hsml, mass, rho, w, neighbours, dx, dy, dwdx, dwdy, dwdr,
sum_CSPM, factor)

      call continuity_density(itimestep, ntotal, mass,dwdx,dwdy, v,x,rho, r, hsml, dx, dy,
dx1, neighbours, w, dt, drhodt)
```

```
call pressure(itimestep, ntotal, x, rho, c_sound, mass,viscosity, p, surface_particles)

call internal_force(itimestep, ntotal, hsml, mass, v, neighbours, r, p, c_sound, pressure_force, viscous_force, rho, viscosity, dx, dy, dwdx, dwdy, dwdr)

call external_force(ntotal,gravity)

call acceleration(ntotal, pressure_force, viscous_force, accel, gravity)

if (XSPH_correction.eq."YES")  call XSPH(ntotal, mass, neighbours, w, rho, v, v_xsph)

call time_integration(accel, ntotal, dt, x, v)

call collision_test(ntotal, radius, itimestep, dt, x, v, n_g, X_0, CR)

call output(ntotal, itimestep, x, v, rho, p, accel, pressure_force, viscous_force, recording_step)

print*, ' '
print*, '-----------------------------------------------------------------------------------'
print*, '  LEAVING THE ITERATION NUMBER ', itimestep, ' OUT OF A TOTAL OF ', maxtimestep
 print*, '---------------------------------------------------------------------------'
 print*, ' '
 print*, ' Simulated physical time (s) = ', i*dt
 print*, ' '

itimestep= itimestep+1

end do
call cpu_time(finish)
time=finish-start
hours_in_seconds=3600;
hours = (time/hours_in_seconds)
minutes = (time -(hours_in_seconds*hours))/60
seconds = (time -(hours_in_seconds*hours)-(minutes*60))

open(unit=807, file='SIMULATION_PARAMETERS.dat')

write(807,*) ' ---------------------------------------------------------------------------------'
write(807,*) ' ----------------------- SIMULATION START --------------------------------'
write(807,*) ' ---------------------------------------------------------------------------------'
write(807,*) '  '
write(807,*) ' TANK LENGTH (m) = ', xlmax
write(807,*) ' '
write(807,*) ' WIDTH HEIGHT IN THE TANK (m) = ', xl
write(807,*) ' '
```

```
   write(807,*) ' WATER HEIGHT IN THE TANK (m) = ' , yl
   write(807,*) ''
   write(807,*) ' The total number of water particles in the domain for this simulation '
   write(807,*)  mp, 'X', np , ' = ', mp*np
   write(807,*) ''
   write(807,*) ' Simulation timestep (s) = ', dt
   write(807,*) ''
   write(807,*) ' Distance between two sucessive centres of mass (m) = ', dx1
   write(807,*) ''
   write(807,*) ' Physical time (s) =', i*dt
   write(807,*) ' '
   write(807,*) ''
   write(807,*) 'Parameter alpha (in the calculation of artificial viscosity) =', alpha
   write(807,*)'  '
   write(807,*) 'CPU Time = '
   write(807,*) hours ,' hours', minutes,' minutes', seconds, ' seconds.'
   write(807,*) ' -------------------------------------------------------------------------------'

   print*,'-------------------------------------------------------------------------------'
   print*, '--------------------------- END OF SIMULATION -------------------------------'
   print*, '-------------------------------------------------------------------------------'
   print*, ' CPU Time = '
   print*, hours ,' hours', minutes,' minutes', seconds, ' seconds.'
   print*, '-------------------------------------------------------------------------------'

   close(445)
   close(450)
   close(455)

  end program main


!----------------------------------------------------------------------------------------------
! Subroutines:
! Files should be created with the name "subroutine_name.f90", for example,
! "geometry.f90"
!----------------------------------------------------------------------------------------------

subroutine geometry (xlmax,ylmax, n_g, X_0)

!--- In this subroutine the planes that define the walls of the reservoir are defined --------

    implicit none

    integer  i,j,active_planes(4)
    double precision  xlmax, ylmax, n_g(4,2), X_0(4,2),summation

! Defining normal vectors to the planes
! Note: if you want to disable a plane, just define the normal vector to !the plane as
n_0=(0.,0.)
```

```
   rewind(445)
   rewind(450)

! The normal vector to the right lateral plane is (-1.,0.) - **plane 01**
   n_g(1,1)=-1.
   n_g(1,2)=0.
! The normal vector to the left lateral plane is (1.,0.) - **plane 02**
   n_g(2,1)=1.
   n_g(2,2)=0.
! The normal vector to the lower plane is (0.,1.) - **plane 03**
   n_g(3,1)=0.
   n_g(3,2)=1.
! The normal vector to the upper plane is (0.,-1.) - **plane 04**  - DISABLED!
   n_g(4,1)=0.
   n_g(4,2)=0.

! Defining a point belonging to each plane
! Note: here you define the geometry of the reservoir

! Point belonging to the right wall
   X_0(1,1)=xlmax
   X_0(1,2)=ylmax

! Point belonging to the left wall
   X_0(2,1)=0.0
   X_0(2,2)=0.0

! Point belonging to the bottom
   X_0(3,1)=xlmax/2.
   X_0(3,2)=0.

! Point belonging to the upper plane
   X_0(4,1)=0.
   X_0(4,2)=0.

   do i=1,4
    write(445,13) (n_g(i,j),j=1,2)
    write(450,13) (X_0(i,j),j=1,2)
13  format(2(2x,F10.3))
   end do

   do i=1,4
    summation=0.
    do j=1,2
     summation = summation + abs(n_g(i,j))
    end do
     active_planes(i)=summation   ! If plane "i" is active, the value of its position in the
vector "active_planes" will be different from zero
   end do
```

```
! Checking active planes

    print*, '------------ Active planes ----------------'
    print*, ' '

    do i=1,4

     if (active_planes(i).ne.0.) then

       if (i.eq.1) then
         print*, '[1] - [RIGHT SIDE]'
         print*, ' '
       else
         if (i.eq.2) then
           print*, '[2] - [LEFT SIDE]'
           print*, ' '
         else
           if (i.eq.3) then
             print*, '[3] - [BOTTOM]'
             print*, ' '
           else
             if (i.eq.4) then
               print*, '[4] - [UPPER PLANE]'
               print*, ' '
             end if
           end if
         end if
       end if

     end if

    end do

end subroutine geometry


subroutine input (ntotal,xl, yl, x,m, mp, np, nx, dx1, kh, mass, rho, hsml, p, viscosity,
K_scale, surface_particles)

!------------------------------------------------------------------------------------------
!  Subroutine for generating initial particle distribution
!  x -- coordinates of particles                                                     [out]
!  mass -- mass of particles                                                         [out]
!  rho -- densities of particles                                                     [out]
!  hsml -- smoothing lengths of particles                                            [out]
!  ntotal -- total domain particle number                                            [out]
!------------------------------------------------------------------------------------------

    implicit none
    include 'param.inc'
```

```
   integer ntotal, i, j, k, nx, m,mp, np, surface_particles(ntotal)
   double precision xl, yl, dx1, dy1, kh, K_scale, max_y_part
   double precision grav, mass(maxn), rho(maxn), hsml(maxn), p(maxn),
viscosity(maxn),x(maxn,2), v(maxn,2)

   grav = 9.81              ! Gravity magnitude
   dx1 = real((xl/nx)/m)    ! Distance between the centre of mass of two adjacent particles
in the x-direction
   dy1 = dx1                ! Distance between the centre of mass of two adjacent particles
in the y-direction
   kh=real((xl/nx)/2.)      ! Cells radius
   K_scale = 2.0            ! Cubic spline kernel

!------------------ Domain Particle Distribution and Initial Properties ------------------------

  open(876, file='../output/data/position/0000000.dat')
  open(877, file='../output/data/velocity/0000000.dat')
  open(879, file='../output/data/pressure/0000000.dat')

   k=0
   do i = 1, mp
   do j = 1, np
    k=k+1
    x(k,1) = (i-1)*dx1 + dx1/2.
    x(k,2) = (j-1)*dy1 + dy1/2.
    v(k,1) = 0.
    v(k,2) = 0.
    rho(k) = 1000.
    mass(k) = dx1*dy1*rho(k)
    viscosity(k)=1.e-3
    hsml(k) = kh/K_scale
    surface_particles(k)=0

    if (x(k,2).gt. max_y_part)   max_y_part = x(k,2)

    write(876,225) k, x(k,1), x(k,2)
    write(877,225) k, v(k,1), v(k,2)

    if ( (j.eq.np).or.(i.eq.mp) ) then    ! Water particles at the free surface
     p(k) = 0.
     surface_particles(k)=k
   end if

 225  format (I8, 2(1x,D21.14))

   end do

  end do

  k=0
```

```
 do i = 1, mp
  do j = 1,np
   k=k+1
   p(k)= rho(k) * grav * (max_y_part -x(k,2))  ! Hydrostatic pressure at the initial instant
   write(879,55)  k, rho (k), mass(k), p(k)
  end do
 end do

55   format(I8,3(1x,D21.14))

 close(876)
 close(877)
 close(879)

 print*,' ***********************************************'
 print*, '     Total number of particles  = ', ntotal
 print*, ' ***********************************************'

 print*, ' Press any key to continue...'
 pause

end subroutine input



subroutine  h_upgrade(dt,ntotal, rho, drhodt, hsml)

!-------------------------------------------------------------------------------------------
!  Subroutine for updating the smoothing radius
!  dt -- time step                                                      [in]
!   ntotal -- number of particles                                       [in]
!   mass  -- particle masses                                            [in]
!   rho -- density                                                      [in]
!   niac -- number of interaction pairs                                 [in]
!   hsml -- smoothing length                                        [in/out]
!   drhodt -- density change rate of each particle                      [in]
!---------------------------------------- -------------------------------------------------

   implicit none

   integer i, ntotal
   double precision dt, hsml(ntotal), rho(ntotal), drhodt(ntotal), dhsmldt(ntotal)

   do i=1, ntotal
        dhsmldt(i) = (-1/2)*(hsml(i)/rho(i))*drhodt(i)
   end do

!-- Integrating and recording the support radius ---

   do i=1,ntotal
```

```
    hsml(i)=hsml(i) + dhsmldt(i)*dt

    if (hsml(i).le.0) hsml(i) = hsml(i) - dt*dhsmldt(i)

   end do

end subroutine h_upgrade


subroutine direct_find(itimestep,ntotal,hsml, neighbours,x,r,dx,dy,K_scale)

!-----------------------------------------------------------------------------------------
!   Subroutine to detect and store neighboring particles.
!   Interaction pairs are determined by directly comparing the particles distance with
!   the corresponding support radius.
!
!   ntotal --  number of particles in the domain                              [in]
!   hsml   --  smoothing length                                               [in]
!   x   --  coordinates of all particles                                      [in]
!  dx -- x-distance between particles i and j                                 [in]
!  dy -- y-distance between particles i and j                                 [in]
!  niac  --  number of neighbours                                             [out]
!  distance --  distance between two particles                                [out]
!  neighbours -- matrix with the neighbors of each particle in the domain     [out]
!-----------------------------------------------------------------------------------------

   implicit none
   include 'param.inc'

   integer  itimestep, ntotal, niac, i, j
   integer  neighbours(maxn,n_neighbours)
   double precision aux, distance, kh, K_scale
   double precision  support_radius(maxn), hsml(maxn), x(maxn,2),
r(maxn,n_neighbours), dx(maxn,n_neighbours), dy(maxn,n_neighbours)

   do i=1,ntotal
     support_radius(i)=K_scale*hsml(i)
   end do

   do i=1,ntotal
     do j=1,n_neighbours
       neighbours(i,j)=0
       dx(i,j)=0.
       dy(i,j)=0.
       r(i,j)=0.
     end do
    end do

    do i=1,ntotal
     niac= 2
```

```
        neighbours(i,1)=i
       do j = 1, ntotal
         aux  = (x(j,1) - x(i,1))**2 + (x(j,2) - x(i,2))**2
         distance = sqrt(aux)

         if (distance.le.support_radius(i)) then
           neighbours(i,niac)=j
           r(i,niac)=distance
           dx(i,niac)=x(j,1) - x(i,1)
           dy(i,niac)=x(j,2) - x(i,2)
           niac=niac+1
         end if

        end do

      end do

      open(222,file='../output/neighbours.dat')    ! Storing the neighbors of each particle in
the domain in the current iteration
      do i=1,ntotal
        write(222,*) (neighbours(i,j), j=1,n_neighbours)
      end do
      close(222)

  end subroutine direct_find



subroutine kernel (ntotal, r, hsml, mass, rho, w, neighbours, dx, dy, dwdx, dwdy, dwdr,
sum_CSPM,factor)

!------------------------------------------------------------------------------------------------------
! Subroutine to calculate the smoothing function (kernel) and its derivatives
!
! if skf  = 1, Cubic spline kernel (defined in param.inc)
!           = 2, Quintic kernel
!           = 3, Quartic kernel by Liu and Lam
! ntotal --  number of particles in the domain                        [in]
! r   -- distance between particles i and j                           [in]
! dx -- x-distance between particles i and j                          [in]
! dy -- y-distance between particles i and j                          [in]
! hsml -- smoothing length                                            [in]
! w – kernel value for all interaction pairs                         [out]
! factor -- kernel's normalisation constant                          [out]
! dwdr -- derivative of kernel in polar coordinates                  [out]
! dwdx -- derivative of kernel with respect to x                     [out]
! dwdy -- derivative of kernel with respect to y                     [out]
!------------------------------------------------------------------------------------------------------

      implicit none
      include 'param.inc'
```

```
integer i, j, ntotal, niac, neighbours(maxn,n_neighbours)
double precision r(maxn,n_neighbours), w(maxn,n_neighbours), dx(maxn,n_neighbours), dy(maxn, n_neighbours), hsml(maxn), dwdx(maxn,n_neighbours), dwdy(maxn,n_neighbours), dwdr(maxn,n_neighbours), mass(maxn), rho(maxn), sum_CSPM(maxn), q, factor

open(88,file='../output/kernel_values.dat')
rewind(88)

do i=1, ntotal
 niac=2
 w(i,1)=0.
 dwdx(i,1)=0.
 dwdy(i,1)=0.
 sum_CSPM(i)=0.      ! To make the CSPM Correction

 do while (neighbours(i,niac).ne.0)
  j=neighbours(i,niac)
  q = r(i,niac)/hsml(i)

  factor = 15./(7.*pi*hsml(i)*hsml(i))      ! Cubic spline kernel
  if (q.ge.0.and.q.le.1.) then
   w(i,niac)= factor* ((2./3.) - q**2. + (1./2.)*q**3)
   dwdr(i,niac) = factor * ((-2.*r(i,niac))/hsml(i)**2 + (3./2.)*(r(i,niac)**2/(hsml(i)**3)) )
   dwdx(i,niac) = factor * ((-2.)/hsml(i)**2 + (3./2.)*(r(i,niac)/hsml(i)**3)) * dx(i,niac)
   dwdy(i,niac) = factor * ((-2.)/hsml(i)**2 + (3./2.)*(r(i,niac)/hsml(i)**3)) * dy(i,niac)

   sum_CSPM(i)= sum_CSPM(i) + (mass(j)/rho(j))*w(i,niac)

   write(88,12) q, w(i,niac)

  else
   if (q.gt.1..and.q.le.2.) then
    w(i,niac) = factor * 1.e0/6.e0 * (2.-q)**3
    dwdr(i,niac) = -factor * (1./(2.*hsml(i))) * ( ( 2. - r(i,niac)/hsml(i))**2. )
    dwdx(i,niac) = -factor * (1./(2.*hsml(i))) * ( ( 2. - r(i,niac)/hsml(i))**2. ) * dx(i,niac)/r(i,niac)
    dwdy(i,niac) = -factor * (1./(2.*hsml(i))) * ( ( 2. - r(i,niac)/hsml(i))**2. ) * dy(i,niac)/r(i,niac)
    sum_CSPM(i)= sum_CSPM(i) + (mass(j)/rho(j))*w(i,niac)

    write(88,12) q, w(i,niac)

   end if
  end if
  niac=niac+1
```

```
      end do ! do while

    end do

12  format(2(1x,D21.14))
    close(88)

end subroutine kernel


subroutine continuity_density(itimestep,ntotal, mass,dwdx,dwdy, v,x,rho, r, hsml, dx, dy,
dx1, neighbours, w, dt, drhodt)

!----------------------------------------------------------------------------------------------------------
!  Subroutine to calculate the density with SPH continuity approach.
!
!  ntotal -- number of particles                                      [in]
!  mass -- particle masses                                            [in]
!  niac  -- number of interaction pairs                               [in]
!  dwdx -- derivation of Kernel for all interaction pairs             [in]
! dwdy -- derivative of kernel with respect to y                      [out]
!  v  -- velocities of all particles                                  [in]
!  x  -- coordinates of all particles                                 [in]
!  rho -- density                                                     [in]
!  drhodt : density change rate of each particle                      [out]
!----------------------------------------------------------------------------------------------------------
    implicit none
    include 'param.inc'

    integer ntotal, niac, itimestep, aux, funit
    double precision mass(maxn), v(maxn,2), x(maxn,2), hsml(maxn), rho(maxn),
drhodt(maxn), w(maxn,n_neighbours), r(maxn,n_neighbours), dx(maxn,n_neighbours),
dy(maxn,n_neighbours), dwdx(maxn,n_neighbours), dwdy(maxn,n_neighbours)
    integer i,j,k,d, neighbours(maxn,n_neighbours)
    double precision    factor, sum_CSPM(maxn), vcc, dv(2), dt, rho0, dx1, dy1,
sum_kernel(maxn)
    character(len= 25) aux2
    character(len=45) name

    do i = 1, ntotal
     drhodt(i) = 0.
    end do

    do k=1,ntotal
      i=neighbours(k,1)
      niac=2

      do while (neighbours(i,niac).ne.0)
        j=neighbours(i,niac)
```

```
      do d=1,2
        dv(d) = v(i,d) - v(j,d)
      end do

        vcc= dv(1)*dwdx(i,niac)
        vcc= vcc+ dv(2)*dwdy(i,niac)
        drhodt(i) = drhodt(i) + mass(j)*vcc
        niac=niac+1

    end do
   end do

   dy1=dx1

!------------------------------------------------------------------------
!   Integrating and recording the density
!------------------------------------------------------------------------

   do k=1,ntotal
    rho(k)=rho(k) + drhodt(k)*dt

    if (mod(itimestep,30).eq.0) mass (k) = rho(k)*dx1*dy1    ! To apply the Sheppard
filter every 30 iterations

   end do

!------------------------------------------------------------------------
!  Applying the Sheppad filter every 30 numerical iterations
!------------------------------------------------------------------------

   if (mod(itimestep,30).eq.0) then

    do k=1,ntotal
      i=neighbours(k,1)
      sum_kernel(k)=0.
      niac=2

     do while (neighbours(i,niac).ne.0)
       j=neighbours(i,niac)
       sum_kernel(k) = sum_kernel(k) + w(i,niac)*mass(j)/rho(j)
       niac=niac+1
     end do

    end do

    do k=1,ntotal
     i=neighbours(k,1)
```

```
        rho(k)=0.
        niac=2

        do while (neighbours(i,niac).ne.0)
         j=neighbours(i,niac)
         rho(k) = rho(k) + mass(j)* w(i,niac)/sum_kernel(k)
         niac=niac+1
        end do

      end do

    end if

end subroutine continuity_density



subroutine pressure(itimestep, ntotal,x,rho,c_sound,mass,viscosity, p, surface_particles)

!-----------------------------------------------------------------------------------------------------
! In this subroutine, the pressure applied to each fluid particle is predicted.
! ntotal -- number of particles                                                    [in]
! rho -- density                                                                   [in]
! p -- pressure                                                                    [out]
! c_sound  -- sound velocity in fluid                                              [out]
!-----------------------------------------------------------------------------------------------------

   implicit none
   include 'param.inc'
   integer i, k, aux, itimestep, ntotal, surface_particles(maxn)
   double precision rho(maxn), mass(maxn), p(maxn),  x(maxn,2), viscosity(maxn)
   double    precision    c_sound,    gamma,    rho0,    delta_x,    min_delta_x,
reference_pressure_part, grav, b, p_tait

!------ Parameters and constants used in Tait's Equation ----
   rho0=1000.     ! density of the fluid at rest
   grav= 9.81     ! gravity magnitude
   gamma=7.
   b=(c_sound**2)*rho0/gamma

   min_delta_x= 100.

   if (itimestep.gt.1) then

     do i =1,ntotal

       do k =1,ntotal

         if (surface_particles(k).ne.0) then      ! particle on the free surface detected
```

```
        !------ We have to find the water particle on the free surface closest to the water
particle being studied (in the x-direction):

           delta_x= abs(x(i,1)-x(surface_particles(k),1))

           if (delta_x.le.min_delta_x) then
            min_delta_x = delta_x

            if      (x(surface_particles(k),2).ge.x(i,2))      reference_pressure_part      =
surface_particles(k)

           end if

          end if

        end do

        aux = rho(i)* grav *(x(reference_pressure_part,2)-x(i,2))
        p_tait=b* (((rho(i)/rho0)**gamma)-1.)
        p(i) = aux + p_tait     ! Hydrostatic pressure + dynamic pressure

      end do

    end if

end subroutine pressure


subroutine internal_force(itimestep, ntotal, hsml, mass, v, neighbours, r, p, c_sound,
pressure_force, viscous_force, rho, viscosity, dx, dy, dwdx, dwdy, dwdr)
!----------------------------------------------------------------------------------------
! Subroutine to calculate the internal forces on the right hand side of the Navier-
! Stokes equations, including the pressure gradient and viscous forces, used by the
! time integration
!
!  itimestep -- current timestep number                                          [in]
!  ntotal -- number of particles                                                 [in]
!  hsml -- smoothing length                                                      [in]
!  mass  -- particle masses                                                      [in]
!  niac   --  number of interaction pairs                                        [in]
!  rho  -- density                                                               [in]
!  viscosity -- dynamic viscosity                                                [in]
!  dwdx -- derivative of kernel with respect to x                                [in]
!  dwdy -- derivative of kernel with respect to y                                [in]
!----------------------------------------------------------------------------------------

    implicit none
    include 'param.inc'
```

```
   integer itimestep, ntotal, niac, i, j, k, d, neighbours(maxn,n_neighbours)
   double precision hsml(maxn), mass(maxn), v(maxn,2), rho(maxn), dwdx(maxn,
n_neighbours), dwdy(maxn, n_neighbours), dwdr(maxn, n_neighbours), r(maxn,
n_neighbours), p(maxn), viscous_force(maxn,2), artificial_viscous_force(maxn,2),
pressure_force(maxn,2), viscosity(maxn), dx(maxn, n_neighbours), dy(maxn,
n_neighbours), den_CSPMx(maxn), den_CSPMy(maxn)
   double precision q, aux, aux_2, aux_3, r2, c_average, amu, pi_visc, c_sound

!-------------------------------------------------------------------------------------------------
!---------------- CALCULATION OF LAMINAR VISCOUS FORCES -------------------
!-------------------------------------------------------------------------------------------------
   do k=1,ntotal
    niac=2
    i = neighbours(k,1)
    viscous_force(i,1)=0.
    viscous_force(i,2)=0.

    do while (neighbours(i,niac).ne.0)
      j = neighbours(i,niac)
      aux = ( (viscosity(j)+viscosity(i)) * mass(j) )/ ( (rho(i)*rho(j)* (r(i,niac)**2) + (0.01
* hsml(i)**2) ) )
      aux_2 = r(i,niac)*dwdr(i,niac) * (v(j,1)-v(i,1))
      aux_3 =  r(i,niac)*dwdr(i,niac) * (v(j,2)-v(i,2))
      viscous_force(i,1)=viscous_force(i,1) + aux * aux_2
      viscous_force(i,2)=viscous_force(i,2) + aux * aux_3
      niac=niac+1
    end do
  end do

!-------------------------------------------------------------------------------------------------
!---------------- CALCULATION OF ARTIFICIAL VISCOUS FORCE ------------------
!-------------------------------------------------------------------------------------------------
   c_average= c_sound

   do k=1,ntotal
     niac=2
     i = neighbours(k,1)
     artificial_viscous_force(i,1)=0.
     artificial_viscous_force(i,2)=0.

    do while (neighbours(i,niac).ne.0)
      j = neighbours(i,niac)
     aux_2 = (v(j,1)-v(i,1))*dx(i,niac) + (v(j,2)-v(i,2))*dy(i,niac)
      r2 = dx(i,niac)**2 + dy(i,niac)**2
      amu = (hsml(i)* aux_2)/ (r2 + 0.01*hsml(i)**2)
      pi_visc = - alpha * c_average * amu / rho(i)
      artificial_viscous_force(i,1) = artificial_viscous_force(i,1) - pi_visc * dwdx(i,niac)
* mass(j)
      artificial_viscous_force(i,2) = artificial_viscous_force(i,2) - pi_visc * dwdy(i,niac)
* mass(j)
      niac=niac+1
```

```
   end do
  end do

!-------------------------------------------------------------------------------------------------------------
!------------------------- CALCULATION OF PRESSURE FORCES -------------------------
!-------------------------------------------------------------------------------------------------------------

   do k=1,ntotal
     niac=2
     i = neighbours(k,1)
     pressure_force(i,1)=0.
     pressure_force(i,2)=0.
     den_CSPMx(i)=0.
     den_CSPMy(i)=0.

     do while (neighbours(i,niac).ne.0)
       j = neighbours(i,niac)
       aux_2 =  - ( (p(i)/rho(i)**2.) + (p(j)/rho(j)**2) ) * dwdx(i,niac) * mass(j)
       aux_3 =  - ( (p(i)/rho(i)**2.) + (p(j)/rho(j)**2) ) * dwdy(i,niac) * mass(j)

       den_CSPMx(i)=  den_CSPMx(i)+ dx(i,niac)* (mass(j)/rho(j)) * dwdx(i,niac)  !
For applying CSPM Correction
       den_CSPMy(i)=  den_CSPMy(i)+ dy(i,niac)* (mass(j)/rho(j)) * dwdy(i,niac)

       pressure_force(i,1)=pressure_force(i,1) +  aux_2
       pressure_force(i,2)=pressure_force(i,2) +  aux_3

       niac=niac+1

     end do
     pressure_force(i,1)=pressure_force(i,1)/den_CSPMx(i)                    +
artificial_viscous_force(i,1)
     pressure_force(i,2)=pressure_force(i,2)/den_CSPMy(i)                    +
artificial_viscous_force(i,2)

   end do

   open(71,file='../output/pressure_force.dat')
   rewind(71)
   write(71,*) itimestep, ' a. iteration'
   do k=1,ntotal
     write(71,*)  pressure_force(i,1), pressure_force(i,2)
   end do

 close (71)

end subroutine internal_force
subroutine external_force (ntotal, gravity)

!-------------------------------------------------------------------------------------------------------------
```

```
!  Subroutine to calculate the external forces, in this case, the gravitational force.
!----------------------------------------------------------------------------------------------------

    implicit none
    include 'param.inc'

    integer i, ntotal
    double precision  gravity(maxn,2)

    do i = 1, ntotal
      gravity(i,1)=0.
      gravity(i,2) = -9.81
    end do

end subroutine external_force



subroutine acceleration (ntotal, pressure_force, viscous_force, accel, gravity)

!----------------------------------------------------------------------------------------------------
!  Subroutine to calculate the particle acceleration
!----------------------------------------------------------------------------------------------------

    implicit none
    include 'param.inc'

    integer i, aux, ntotal
    double precision pressure_force(maxn,2), viscous_force(maxn,2), gravity(maxn,2),
accel(maxn,2)

    do i =1, ntotal
      accel(i,1) = -pressure_force(i,1) + viscous_force(i,1)+gravity(i,1)
      accel(i,2) = -pressure_force(i,2) + viscous_force(i,2)+gravity(i,2)
    end do

end subroutine acceleration



subroutine XSPH(ntotal, mass, neighbours, w, rho, v, v_xsph)

!----------------------------------------------------------------------------------------------------
!  Subroutine to calculate the average velocity to correct velocity for preventing
!  penetration (Monaghan, 1992)
!
!  ntotal –  number of particles                                   [in]
!  mass -- particle masses                                         [in]
!  w -- Kernel for all interaction pairs                           [in]
!  v -- velocity of each particle                                  [in]
!  v_xsph -- average velocity (XSPH)                               [out]
```

```
!   rho -- density                                                  [in]
!----------------------------------------------------------------------

   implicit none
   include 'param.inc'

   integer ntotal,niac,neighbours(maxn,n_neighbours)
   double precision mass(maxn), w(maxn,n_neighbours), rho(maxn), v(maxn,2),
v_xsph(maxn,2), rho_average
   integer i,j,k,d
   double precision dv(2), epsilon

!  epsilon --- a small constants chosen by experience, may lead to instability.

   epsilon = 0.50    ! 0.50 is a tipycal value

   do i = 1, ntotal
    do d = 1, 2
      v_xsph(i,d) = 0.
    end do
   end do

   do k=1,ntotal
    niac = 2
    i = neighbours(k,1)

    do while (neighbours(i,niac).ne.0)
     j=neighbours(i,niac)
    do d=1,2
      rho_average = (rho(i)+rho(j))/2.
      dv(d) = v(j,d) - v(i,d)
       v_xsph (i,d) =  v_xsph(i,d)  + mass(j)*dv(d)*w(i,niac)/rho_average
    end do
    niac=niac+1

   end do

  end do

  do k = 1, ntotal
   do d = 1, 2
     v_xsph(i,d) =  epsilon * v_xsph (i,d)
   end do
  end do

end subroutine XSPH


subroutine time_integration (accel, ntotal, dt, x, v, v_xsph)
!----------------------------------------------------------------------
```

```
!  x -- coordinates of particles                                    [in/out]
!  v -- velocities of particles                                     [in/out]
!  dt -- timestep                                                       [in]
!-----------------------------------------------------------------------------

    implicit none
    include 'param.inc'

    integer ntotal, i, k
    double precision x(maxn,2), v(maxn,2), accel(maxn,2), v_xsph(maxn,2), dt

!-----------------------------------------------------------------------------
!  Files used to store particle positions and velocities at the beginning of the numerical
!  iteration  (before temporal integration)
!-----------------------------------------------------------------------------
    open(unit=40,file='../output/initial_velocity.dat')   ! particles velocities
    open(unit=43,file='../output/initial_position.dat')   ! particles positions
!-----------------------------------------------------------------------------

    rewind(40)
    rewind(43)

    do i=1, ntotal
      write(43,31) (x(i,k), k=1,2)
      write(40,31) (v(i,k), k=1,2)
    end do
 31  format(2(1x,D21.14))

  close(40)
  close(43)

!-----------------------------------------------------------------------------
!   Performing temporal integration
!   (Euler integration method (efficient for small time steps))
!-----------------------------------------------------------------------------
    do i=1, ntotal
      x(i,1) = x(i,1) + v(i,1)*dt
      x(i,2) = x(i,2) + v(i,2)*dt

      if (XSPH_correction.eq. 'NO') then
        v(i,1) = v(i,1) + accel(i,1)*dt
        v(i,2) = v(i,2) + accel(i,2)*dt
      else
        if (XSPH_correction.eq. 'YES')  then
          v(i,1) = v(i,1) + accel(i,1)*dt  + v_xsph(i,1)
          v(i,2) = v(i,2) + accel(i,2)*dt + v_xsph(i,2)
        end if
      end if
    end do
```

```
end subroutine time_integration


subroutine collision_test(ntotal, radius, itimestep, dt, x, v, n_g, X_0, CR)

!-----------------------------------------------------------------------------------------------
! In this subroutine, particle collisions against the boundaries are detected and reflections
! towards the interior of the domain are performed.
!
! ntotal -- number of particles                                    [in]
! radius -- particles' radius                                      [in]
! dt -- timestep                                                   [in]
! x -- coordinates of particles                                    [in/out]
! v -- velocities of particles                                     [in/out]
! n_g -- normal vector to plane                                    [in]
! X_0 -- point belonging to the plane                              [in]
! CR -- coefficient of restitution of kinetic energy               [in]
!-----------------------------------------------------------------------------------------------

   implicit none
   include 'param.inc'

   integer itimestep,ntotal
   double precision radius,dt, CR
   integer n_part,i,j,aux,ncol,plane_col,plane_col2,plane,cont,d
   double precision aux_x, aux_y, dx, dy, module_a, num, den, aux1_s, aux2_s,
esc_1,esc_2,esc_3,esc_4,esc_5,num_escalar,den_escalar,minor_d,minor_d2
   double precision x(maxn,2), v(maxn,2), P_0(4,3), P_I(4,3), a(2), s(4),
points_distance(4), X_0(4,2), n_g(4,2), pos_reflected(2), new_C0(2)
   integer, dimension(:,:), allocatable:: planes
   double precision, dimension (:,:), allocatable:: position_t0, position_t1, velocity_t1,
distance, xp_0
   character(len=3) stop, corner_reflection
   character(len=20) planecol, planecol2

   allocate (position_t0(maxn,2), position_t1(maxn,2), velocity_t1(maxn,2),
distance(maxn,5), planes(maxn,5) ,xp_0(maxn,2))

!---------------------Input files---------------------

   open(unit=43,file='../output/initial_position.dat') ! particle positions at the beginning
of this iteration

   rewind(43)
   rewind(445)  !vectors normal to the reservoir wall
   rewind(450)  !points belonging to the reservoir walls

   do n_part=1,ntotal
    read(43,*) (position_t0(n_part,j),j=1,2)
    xp_0(n_part,1) = position_t0(n_part,1)
    xp_0(n_part,2) = position_t0(n_part,2)
```

```
  end do
  close(43)

!--- particle positions and velocities at the end of the iteration assuming that the particles
did not collide against the planes

   do n_part=1,ntotal
     position_t1(n_part,1)= x(n_part,1)
     position_t1(n_part,2)= x(n_part,2)
     velocity_t1(n_part,1)= v(n_part,1)
     velocity_t1(n_part,2)= v(n_part,2)
   end do

!*******************************************************************
! Calculating the distance between the center of the particle in the initial position "C0"
! and in the final position "C1" and the planes
!*******************************************************************

   stop="NO"      !---------------- Assumption that there were collisions in the first
numerical iteration

   do n_part=1,ntotal       ! Loop for all particles
    cont=0
    distance(n_part,1) = n_part

!*******************************************************************
! Checking the possibilities of collisions for the particle "n_part"
!*******************************************************************
    do while (stop.eq."NO")  !AAA

!--- Calculations of the distance from the center "C1" of the particle  "n_part" at the end
! point of its trajectory, after each possible collision, and each active plane:

 !--- Distance to plane 01 (column 02 of the distance matrix):

      if (n_g(1,1).ne.0.or.n_g(1,2).ne.0) then
       aux_x=n_g(1,1)*(position_t1(n_part,1)-X_0(1,1))
       aux_y=n_g(1,2)*(position_t1(n_part,2)-X_0(1,2))
       distance(n_part,2)=aux_x+aux_y

      ! To ensure accuracy in collision detection:
       distance(n_part,2) = ANINT(distance(n_part,2)*1000000.0)/1000000.0
      else
       distance(n_part,2)=99999.
      end if

!--- Distance to plane 02 (column 03 of the distance matrix):

      if (n_g(2,1).ne.0.or.n_g(2,2).ne.0) then
       aux_x=n_g(2,1)*(position_t1(n_part,1)-X_0(2,1))
```

```
      aux_y=n_g(2,2)*(position_t1(n_part,2)-X_0(2,2))

      distance(n_part,3)=aux_x+aux_y

     ! To ensure accuracy in collision detection:
      distance(n_part,3) = ANINT(distance(n_part,3)*1000000.0)/1000000.0
    else
      distance(n_part,3)=99999.
    end if

!--- Distance to plane 03 (column 04 of the distance matrix):

    if (n_g(3,1).ne.0.or.n_g(3,2).ne.0) then
      aux_x=n_g(3,1)*(position_t1(n_part,1)-X_0(3,1))
      aux_y=n_g(3,2)*(position_t1(n_part,2)-X_0(3,2))

      distance(n_part,4)=aux_x+aux_y

     ! To ensure accuracy in collision detection:
      distance(n_part,4) = ANINT(distance(n_part,4)*1000000.0)/1000000.0

    else
      distance(n_part,4)=99999.
    end if

!--- Distance to plane 04 (column 05 of the distance matrix):

    if (n_g(4,1).ne.0.or.n_g(4,2).ne.0) then
      aux_x=n_g(4,1)*(position_t1(n_part,1)-X_0(4,1))
      aux_y=n_g(4,2)*(position_t1(n_part,2)-X_0(4,2))
      distance(n_part,5)=aux_x+aux_y

     ! To ensure accuracy in collision detection:
      distance(n_part,5) = ANINT(distance(n_part,5)*1000000.0)/1000000.0
    else
      distance(n_part,5)=99999.
    end if
! ---- Initializing the "planes" matrix with zeros:

    planes(n_part,1)=n_part
    do j=2,5
      planes(n_part,j)=0  ! Assuming initially that there is no collision against any plane
    end do

!*****************************************************************
! Performing the collision test, detecting and storing the planes against which collisions
! may have occurred
! Criterion: distances from the center of the particle at t0+dt and the plane considered
!*****************************************************************

    do j=2,5
```

```
! If there is a possibility of collision against any plane, the position of the "planes" array
! relative to the respective collision plane receives the value "one"

      if (distance(n_part,j).ge.0.and.distance(n_part,j).le.radius) then
        planes(n_part,j)=1
      else
       if (distance(n_part,j).lt.0) then
         planes(n_part,j)=1
       end if
      end if

     end do

!**********************************************************************
! If there was a collision of the particle, it is necessary to check if it will still continue to
! collide with the planes within the planes within the time step
! *********************************************************************
      stop="YES"   ! Assumpting there was no collision
     j=2
     do while (stop.eq."YES".and.j.LE.5)
       if (planes(n_part,j).eq.1) then   ! If true, need to check for new collisions
         if (stop.eq."YES") then
           cont=cont+1          ! Record the possible number of collisions
           stop="NO"
         end if
       else
        j=j+1
       end if
     end do

!**********************************************************************
! Calculation of "P_0" points of the "n_part" particle in relation to each possible
! collision plane
! Calculations are performed only if there is a possibility of collision of the particles
! against one or more planes
!**********************************************************************

     if (stop.eq."NO") then        !!!!B

      open(unit=773,file='../output/P_0.dat') ! file containing the coordinates of point
P_0 (closest to the collision plane)
      rewind(773)
      write (773,*) n_part
      do i=1,4      ! Initializing the matrix P_0 (referring to the particle "n_part")
        P_0(i,1)=i
        do j=2,3
          P_0(i,j)=0.
        end do     !Columns 2 to 3 of the matrix are set to "zero"
```

```
      end do    !It is assumed that there was no collision against any plane

!*************************************************************************
! Determining P_0 (position of the particle point that will touch each of the possible
! collision planes).
! Only if there is a possibility of collision
!*************************************************************************

      do i=2,5
        aux=planes(n_part,i)
        if (aux.ne.0) then
          if (i.eq.2) then      ! P_0 relative to plane 01 [Right side]
            P_0(i-1,2)=position_t0(n_part,1)+radius
            P_0(i-1,3)=position_t0(n_part,2)
          else
            if (i.eq.3) then    !P_0 relative to plane 02 [Left side]
              P_0(i-1,2)=position_t0(n_part,1)-radius
              P_0(i-1,3)=position_t0(n_part,2)
            else
              if (i.eq.4) then    !P_0 relative to plane 03  [Bottom]
                P_0(i-1,2)=position_t0(n_part,1)
                P_0(i-1,3)=position_t0(n_part,2)-radius
              else
                if (i.eq.5) then   !P_0 relative to plane 04  [Upper]
                  P_0(i-1,2)=position_t0(n_part,1)
                  P_0(i-1,3)=position_t0(n_part,2)+radius
                end if
              end if
            end if
          end if
        end if
      end do
      do i=1,4
        write(773,27) (P_0(i,j),j=1,3) ! Writing P_0 to an output file
27      format (3(2x,E13.6))
      end do
      close(773)

!*************************************************************************
! Finding the directing vector "a" of the line cointaining the trajectory of "P_0"
! Calculate "a" from the positions of the centers of the particles at the beginning (C0) and
! at the end of the trajectory (C1)
! It is known that the lines containing points P_0 and P_1 (line 1) and the one containing
! points C0 and C1 are parallel
!*************************************************************************

      dx=position_t1(n_part,1)-position_t0(n_part,1)
      dy=position_t1(n_part,2)-position_t0(n_part,2)

      module_a=sqrt(dx*dx+dy*dy)
```

```
! Obtaining the directing vector of the line that contains the trajectory of the particle until
the collision

        do i=1,2

         if ((module_a).ne.0) then    ! The particle moved
          a(i)=(position_t1(n_part,i)-position_t0(n_part,i))/module_a
         else                 ! The particle is still
          a(i) = 0.
         end if

        end do

!*********************************************************************
! Calculating the parameters of the line that intercepts the collision plane - equation of the
! line: P_I = P0 + s*a
!*********************************************************************
        do i=1,4
         s(i)=99999.    ! If there is no collision against the plane i, the value of 's' relative to
the collision against that plane will be 99999.
        end do

        do i=1,4
          if ((i.eq.1).and.(planes(n_part,i+1).ne.0)) then

       !*****Possible collision against the Plane 01 ***********************

       ! X_0 belongs to the plane and P_0 belongs to the line
       ! Calculating X_0-P_0:
            aux1_s=X_0(1,1)-P_0(i,2)
            aux2_s=X_0(1,2)-P_0(i,3)

       ! Taking the scalar product between (X_0-P_0) and the normal vector to the plane:
            esc_1=(aux1_s)*n_g(1,1)
            esc_2=(aux2_s)*n_g(1,2)
            num_escalar=esc_1+esc_2

       ! Making the scalar product between the direction vector of the line and the normal
vector to the plane:
            esc_4=a(1)*n_g(1,1)
            esc_5=a(2)*n_g(1,2)
            den_escalar=esc_4+esc_5

            if (den_escalar.ne.0) then
             s(1)=num_escalar/den_escalar
            else
              s(1) = 0.
            end if
```

```
!---------- Collision point against the plane 01 ---------------------
     P_I(i,1)=i
     P_I(i,2) = P_0(i,2) + s(1)*a(1)
     P_I(i,3) = P_0(i,3) + s(1)*a(2)

     else
      if ((i.eq.2).and.(planes(n_part,i+1).ne.0)) then

!******Possible collision com o plane 02***********************

        aux1_s=X_0(2,1)-P_0(i,2)
        aux2_s=X_0(2,2)-P_0(i,3)

        esc_1=(aux1_s)*n_g(2,1)
        esc_2=(aux2_s)*n_g(2,2)

        num_escalar=esc_1+esc_2

        esc_4=a(1)*n_g(2,1)
        esc_5=a(2)*n_g(2,2)

        den_escalar=esc_4+esc_5

        if (den_escalar.ne.0) then
          s(2)=num_escalar/den_escalar
        else
          s(2) = 0.
        end if

!---------- Collision point against the plane 02---------------------
        P_I(i,1)=i
        P_I(i,2) = P_0(i,2) + s(2)*a(1)
        P_I(i,3) = P_0(i,3) + s(2)*a(2)

        else
          if ((i.eq.3).and.(planes(n_part,i+1).ne.0)) then

!*********Possible collision against the plane 03*****************
          aux1_s=X_0(3,1)-P_0(i,2)
          aux2_s=X_0(3,2)-P_0(i,3)

          esc_1=(aux1_s)*n_g(3,1)
          esc_2=(aux2_s)*n_g(3,2)

          num_escalar=esc_1+esc_2

          esc_4=a(1)*n_g(3,1)
          esc_5=a(2)*n_g(3,2)
          den_escalar=esc_4+esc_5
```

```
            if (den_escalar.NE.0) then
              s(3)=num_escalar/den_escalar
            else
              s(3) = 0.
            end if
!---------- Collision point against the plane 03--------------------
            P_I(i,1)=i
            P_I(i,2) = P_0(i,2) + s(3)*a(1)
            P_I(i,3) = P_0(i,3) + s(3)*a(2)

          else
            if ((i.eq.4).and.(planes(n_part,i+1).ne.0)) then

!******** Possible collision against the plane 04**************

              aux1_s=X_0(4,1)-P_0(i,2)
              aux2_s=X_0(4,2)-P_0(i,3)

              esc_1=(aux1_s)*n_g(4,1)
              esc_2=(aux2_s)*n_g(4,2)

              num_escalar=esc_1+esc_2

              esc_4=a(1)*n_g(4,1)
              esc_5=a(2)*n_g(4,2)

              den_escalar=esc_4+esc_5

              if (den_escalar.NE.0) then
                s(4)=num_escalar/den_escalar
              else
                s(4) = 0.
              end if

!---------- Collision point against the plane 04 ----------------
              P_I(i,1)=i
              P_I(i,2) = P_0(i,2) + s(4)*a(1)
              P_I(i,3) = P_0(i,3) + s(4)*a(2)

            end if
          end if
        end if
      end if
    end do

!*********************************************************************
! Calculating (P_I-P_0), in absolute value, for possible collision planes
! (Checking between all possible collision planes)
!*********************************************************************
```

```
      do i=1,4
        points_distance(i)=abs(s(i))     ! --- The distance between points P_0 and P_I is
equal to the absolute value of s(i)
      end do

!***********************************************************************
! Checking the shortest distance between points on the trajectory to the collision and
! determining the collision plane(s)
!***********************************************************************

      stop="YES"     ! Assumpting there was no collision

      minor_d=99999.
      minor_d2=99999.
      ncol=0
      plane_col=0     ! Assuming there was no collision against any plane
      plane_col2=0    ! There may be simultaneous collision between two planes (corner)

      do i=1,4     ! In this loop, the collision planes are identified
       if (points_distance(i).lt.minor_d) then
         ncol=1
         plane_col=i     !If there was a collision, storing the collision plane
         minor_d= points_distance(i)
       else
         if ((ncol.eq.1).and.( points_distance(i).eq.minor_d)) then  ! If there is a collision
at the corner (meeting between two planes)
           ncol=ncol+1
           plane_col2=i
           minor_d2= points_distance(i)
         end if
       end if
     end do

     if (ncol.ne.0) then
      if (minor_d.lt.minor_d2) then
       ncol=1
      else
       if (minor_d2.lt.minor_d) then
         ncol=1
         plane_col=plane_col2
         minor_d=minor_d2
       else     ! Collision of the particle with two planes simultaneously
         if (minor_d.eq.minor_d2) then
           ncol=2
         end if
       end if
      end if
     end if

!***********************************************************************
! Just checking the collision of the particle "n_part"
```

```
!***********************************************************************

      if (ncol.ne.0) then  !!!E

       if (ncol.eq.1) then    !!!F     ! If there was a collision against a plane, the particle
position and velocity must be updated.
        stop="NO"

        if (plane_col.eq.1) then
         planecol="[1] - [RIGHT PLANE]"
        else
         if (plane_col.eq.2) then
          planecol="[2] - [LEFT PLANE]"
         else
          if (plane_col.eq.3) then
           planecol="[3] - [BOTTOM]"
          else
           if (plane_col.eq.4) then
            planecol= "[4] - [UPPER PLANE]"
           end if
          end if
         end if
        end if

!***********************************************************************
! PERFORMING THE REFLECTION OF THE PARTICLE "n_part"
! Reflecting the final position of the center to the interior of the domain
!***********************************************************************
        if (plane_col.eq.1) then  ! collision against the plane 01

         if (CR.gt.0.) then

          pos_reflected(1)=position_t1(n_part,1)+((1.     +     CR     )*(radius-
distance(n_part,2)))*n_g(1,1)
          pos_reflected(2)=position_t1(n_part,2)+(2*(radius-
distance(n_part,2)))*n_g(1,2)

! Updating the velocity after each collision
          velocity_t1(n_part,1)=(-1)*CR*velocity_t1(n_part,1)  !x-direction

         else

          pos_reflected(1)=P_I(1,2)-radius
          pos_reflected(2)=P_I(1,3)
          velocity_t1(n_part,1)=(-1.)*CR*velocity_t1(n_part,1)  !x-direction

         end if

        else
```

```
    if (plane_col.eq.2) then  ! collision against the plane 02

     if (CR.gt.0.) then
       pos_reflected(1)=position_t1(n_part,1)+((1.      +      CR      )*(radius-
distance(n_part,3)))*n_g(2,1)
       pos_reflected(2)=position_t1(n_part,2)+(2*(radius-
distance(n_part,3)))*n_g(2,2)

     ! Updating the velocity after each collision
       velocity_t1(n_part,1)=(-1)*CR*velocity_t1(n_part,1)   !x-direction

     else
       pos_reflected(1)=P_I(2,2)+radius   ! position of the center of mass
       pos_reflected(2)=P_I(2,3)

       velocity_t1(n_part,1)=(-1.)*CR*velocity_t1(n_part,1)   !x-direction
     end if
   else

    if (plane_col.eq.3) then  ! collision against the plane 03

     if (CR.gt.0.) then
       pos_reflected(1)=position_t1(n_part,1)+(2*(radius-
distance(n_part,4)))*n_g(3,1)
       pos_reflected(2)=position_t1(n_part,2)+((1.      +      CR      )*(radius-
distance(n_part,4)))*n_g(3,2)

       velocity_t1(n_part,2)=(-1)*CR*velocity_t1(n_part,2)  !y-direction

     else
       pos_reflected(1)=P_I(3,2)
       pos_reflected(2)=P_I(3,3)+radius

       velocity_t1(n_part,2)=(-1.)*CR*velocity_t1(n_part,2)  !y-direction
     end if

    else

     if (plane_col.eq.4) then  ! collision against the plane 04

       if (CR.gt.0.) then

         pos_reflected(1)=position_t1(n_part,1)+(2*(radius-
distance(n_part,5)))*n_g(4,1)
         pos_reflected(2)=position_t1(n_part,2)+((1.      +      CR      )*(radius-
distance(n_part,5)))*n_g(4,2)

         velocity_t1(n_part,2)=(-1)*CR*velocity_t1(n_part,2)  !y-direction
       else
         pos_reflected(1)=P_I(4,2)
```

```
          pos_reflected(2)=P_I(4,3)-radius

          velocity_t1(n_part,2)=(-1.)*CR*velocity_t1(n_part,2)   !y-direction

        end if

      end if
    end if
  end if
end if

!*********************************************************************
! Obtaining the new position of the center of the particle "new C_0" for !the new trajectory
! after the previous collisions
! This is the position of the mass center of the particle at the instant of the previous
! collision against the plane
!*********************************************************************
   new_C0(1)=position_t0(n_part,1)+s(plane_col)*a(1)
   new_C0(2)=position_t0(n_part,2)+s(plane_col)*a(2)

!*********************************************************************
! Updating the initial position matrix of the particle "n_part". This is the position of the
! particle center for the next iteration
!*********************************************************************
   do j=1,2
    position_t0(n_part,j)=new_C0(j)
   end do

!*********************************************************************
! Updating the matrix with the final position of the particle "n_part" !from the coordinates
! after reflection
! This is the final position of the particle mass centre after the !collision. If there are no
! more collisions, its position is already !defined.
!*********************************************************************
   do j=1,2
    position_t1(n_part,j)=pos_reflected(j)
   end do

!*********************************************************************
!*********** COLLISION AGAINST TWO PLANES SIMULTANEOUSLY
!*********************************************************************

    else   !!!F
     if (ncol.eq.2) then
     stop="NO"
     corner_reflection="NO"

      if (plane_col.eq.1) then
       planecol="[1] - [RIGHT PLANE]"
      else
```

```
      if (plane_col.eq.2) then
       planecol="[2] - [LEFT PLANE]"
      else
       if (plane_col.eq.3) then
        planecol="[3] - [BOTTOM] "
       else
        if (plane_col.eq.4) then
         planecol='[4] - [UPPER PLANE]'
        end if
       end if
      end if
     end if

     if (plane_col2.eq.1) then
      planecol2="[1] - [RIGHT PLANE]"
     else
      if (plane_col2.EQ.2) then
       planecol2="[2] - [LEFT PLANE]"
      else
       if (plane_col2.EQ.3) then
        planecol2="[3] - [BOTTOM] "
       else
        if (plane_col2.EQ.4) then
         planecol2='[4] - [UPPER PLANE]'
        end if
       end if
      end if
     end if

!*****************************************************************
! Performing the reflection of the particle "n_part"
! Reflecting the final position of the center to the interior of the !domain
!*****************************************************************

!-------------- Collision against the first plane -----------------------------------------------

    if (plane_col.eq.1) then   ! Collision against the plane 01

     if (CR.gt.0.) then
     ! Reflecting the final position of the center of mass into the interior of the domain
      pos_reflected(1)=position_t1(n_part,1)+((1.     +     CR     )*(radius-
distance(n_part,2)))*n_g(1,1)
      pos_reflected(2)=position_t1(n_part,2)+(2*(radius-
distance(n_part,2)))*n_g(1,2)
     else
      pos_reflected(1)=P_I(1,2)-radius
      pos_reflected(2)=P_I(1,3)
     end if
!*****************************************************************
! Correcting particle velocity and updating velocity matrix after each !collision
```

```
!***************************************************************************
        velocity_t1(n_part,1)=(-1.)*CR*velocity_t1(n_part,1)

       else
        if (plane_col.eq.2) then ! Collision against the plane 02

         if (CR.gt.0.) then
          pos_reflected(1)=position_t1(n_part,1)+((1.        +        CR        )*(radius-
distance(n_part,3)))*n_g(2,1)
          pos_reflected(2)=position_t1(n_part,2)+(2*(radius-
distance(n_part,3)))*n_g(2,2)
         else
          pos_reflected(1)=P_I(2,2)+radius
          pos_reflected(2)=P_I(2,3)
         end if

       ! Updating the velocity at the end of the iteration
         velocity_t1(n_part,1)=(-1.)*CR*velocity_t1(n_part,1)
        else
         if (plane_col.eq.3) then  ! Collision against the plane 03

          if (CR.gt.0.) then
            pos_reflected(1)=position_t1(n_part,1)+(2*(radius-
distance(n_part,4)))*n_g(3,1)
            pos_reflected(2)=position_t1(n_part,2)+((1.        +        CR        )*(radius-
distance(n_part,4)))*n_g(3,2)
          else
           pos_reflected(1)=P_I(3,2)
           pos_reflected(2)=P_I(3,3)+radius
          end if

        ! Updating the velocity at the end of the iteration
         velocity_t1(n_part,2)=(-1)*CR*velocity_t1(n_part,2)

         else

          if (plane_col.eq.4) then  ! Collision against the plane 04

           if (CR.gt.0.) then
             pos_reflected(1)=position_t1(n_part,1)+(2*(radius-
distance(n_part,5)))*n_g(4,1)
             pos_reflected(2)=position_t1(n_part,2)+((1.        +        CR        )*(radius-
distance(n_part,5)))*n_g(4,2)
           else
            pos_reflected(1)=P_I(4,2)
            pos_reflected(2)=P_I(4,3)-radius
           end if

          ! Updating the velocity at the end of the iteration
           velocity_t1(n_part,2)=(-1)*CR*velocity_t1(n_part,2)
```

```
        end if
      end if
    end if
  end if

!**********************************************************************
! Updating the final position matrix of the particle "n_part" from the coordinates after the
! first reflection
!**********************************************************************
  do j=1,2
   position_t1(n_part,j)=pos_reflected(j)
  end do

!--------------- Collision against the second plane ------------------------------------------------

  if (plane_col2.eq.1) then  ! Collision against the plane 01

   if (CR.gt.0.) then
    pos_reflected(1)=position_t1(n_part,1)+((1.        +          CR          )*(radius-
distance(n_part,2)))*n_g(1,1)
    pos_reflected(2)=position_t1(n_part,2)+(2*(radius-distance(n_part,2)))*n_g(1,2)
   else
    pos_reflected(1)=P_I(1,2)-radius
    pos_reflected(2)=P_I(1,3)
   end if

!**********************************************************************
! Correcting particle velocity and updating velocity matrix after each !collision
!**********************************************************************

   velocity_t1(n_part,1)=(-1.)*CR*velocity_t1(n_part,1)

  else
   if (plane_col2.eq.2) then ! Collision against the plane 02

    if (CR.gt.0.) then
     pos_reflected(1)=position_t1(n_part,1)+((1.         +          CR          )*(radius-
distance(n_part,3)))*n_g(2,1)
     pos_reflected(2)=position_t1(n_part,2)+(2*(radius-
distance(n_part,3)))*n_g(2,2)
    else
     pos_reflected(1)=P_I(2,2)+radius
     pos_reflected(2)=P_I(2,3)
    end if

!**********************************************************************
!--------------- Updating velocity matrix
!**********************************************************************
```

```
      velocity_t1(n_part,1)=(-1.)*CR*velocity_t1(n_part,1)
       else
        if (plane_col2.eq.3) then  ! Collision against the plane 03

          if (CR.gt.0.) then
            pos_reflected(1)=position_t1(n_part,1)+(2*(radius-
distance(n_part,4)))*n_g(3,1)
            pos_reflected(2)=position_t1(n_part,2)+((1.     +      CR      )*(radius-
distance(n_part,4)))*n_g(3,2)

        else
          pos_reflected(1)=P_I(3,2)
          pos_reflected(2)=P_I(3,3)+radius
        end if

!*******************************************************************
!--------------- Updating velocity matrix
!*******************************************************************
        velocity_t1(n_part,2)=(-1.)*CR*velocity_t1(n_part,2)

        else
         if (plane_col2.eq.4) then  ! Collision against the plane 04

          if (CR.gt.0.) then
           pos_reflected(1)=position_t1(n_part,1)+(2*(radius-
distance(n_part,5)))*n_g(4,1)
            pos_reflected(2)=position_t1(n_part,2)+((1.     +      CR      )*(radius-
distance(n_part,5)))*n_g(4,2)

          else
            pos_reflected(1)=P_I(4,2)
            pos_reflected(2)=P_I(4,3)-radius
          end if

!*******************************************************************
! --------------- Updating velocity matrix
!******************************************************************

          velocity_t1(n_part,2)=(-1.)*CR*velocity_t1(n_part,2)

         end if
       end if
      end if
     end if

!*******************************************************************
! The next commands are used to remove the infinite loop if the reflection
! of the particle puts it in contact with a corner (encounter between 2 planes)
******************************************************************
     if ( (distance(n_part,2).eq.radius).and.(distance(n_part,4).eq.radius)) then
```

```
corner_reflection = "YES"
else
 if ( (distance(n_part,2).eq.radius).and.(distance(n_part,5).eq.radius)) then
  corner_reflection = "YES"
 else
   if ( (distance(n_part,3).eq.radius).and.(distance(n_part,4).eq.radius)) then
  corner_reflection = "YES"
   else
     if ( (distance(n_part,3).eq.radius).and.(distance(n_part,5).eq.radius)) then
      corner_reflection = "YES"
     end if
   end if
 end if
end if

!-----------------------------------------------------------------------------------------
! Finding the position of the particle center at the moment of collision
!-----------------------------------------------------------------------------------------

do i=1,2
  new_C0(i)=position_t0(n_part,i)
end do

if ((plane_col.eq.1).OR.(plane_col.eq.2)) then
  new_C0(1)=position_t0(n_part,1)+s(plane_col)*a(1)
else
  if ((plane_col.eq.3).OR.(plane_col.eq.4)) then
    new_C0(2)=position_t0(n_part,2)+s(plane_col)*a(2)
  end if
end if

if ((plane_col2.eq.1).OR.(plane_col2.eq.2)) then
  new_C0(1)=position_t0(n_part,1)+s(plane_col2)*a(1)
else
  if ((plane_col2.eq.3).OR.(plane_col2.eq.4)) then
    new_C0(2)=position_t0(n_part,2)+s(plane_col2)*a(2)
  end if
end if

!*******************************************************************
! Updating the initial position matrix of the particle "n_part". This is the position of the
! particle centre  for the next iteration
!*******************************************************************

do j=1,2
  position_t0(n_part,j)=new_C0(j)
end do

!*******************************************************************
! Updating the final position matrix of the particle "n_part" from the coordinates after the
! second reflection
```

```
!*******************************************************************
      do j=1,2
        position_t1(n_part,j)=pos_reflected(j)
      end do

      end if  ! collision against 02 planes
     end if   ! collision against only 01 plane      !!!F

  end if  !!!E

  else  !!!!B   ! It will only enter this ELSE if there is no more collision (distances to
planes greater than zero)

   stop="YES"

  end if !!!!! B

  if (corner_reflection.eq."YES") then
    stop = "YES"
  end if

 ! Stopping criterion: if the particle had contacted with the wall and is at rest

  if ( (velocity_t1(n_part,1).eq.0.).and.(velocity_t1(n_part,2).eq.0.) ) then
    stop = "YES"
  end if

end do !  AAA do while

stop="NO"

!-----------------------------------------------------------------------------------------
! Great! All collisions for the particle "n_part"  have been performed at the current time
! step. Now, let's change the particle
!-----------------------------------------------------------------------------------------

end do  ! Particle change

write(455,*) '-------------------------------------------------------------------'
write(455,*) '     FINAL RESULTS FOR THE ITERATION ',  itimestep
write(455,*) '-------------------------------------------------------------------'
do i=1,ntotal
write(455,*) 'Final reflected position of the center of mass (Cf):'
write(455,*) i, (position_t1(i,j),j=1,2)
write(455,*) ' '
end do

do i=1,ntotal
  write(455,*) 'Final velocity of the centre of mass (Vf):'
  write(455,*) i, (velocity_t1(i,j),j=1,2)
```

```
  write(455,*) ' '
end do

write(455,*) '-------------------------------------------------------------'
write(455,*) ' '

! Updating positions and velocities for the next iteration:

do i=1,ntotal
  x(i,1) = position_t1(i,1)
  x(i,2) = position_t1(i,2)
  v(i,1) = velocity_t1(i,1)
  v(i,2) = velocity_t1(i,2)

end do

end subroutine collision_test


subroutine output(ntotal, itimestep, x, v, rho, p, accel, pressure_force, viscous_force,
recording_step)

!---------------------------------------------------------------------------------------------
!   Subroutine for saving particles physical properties to files
!
! ntotal--  number of particles                                                        [in]
!   x-- coordinates of particles                                                       [in]
!   v-- velocities of particles                                                        [in]
!   accel-- accelerations of particles                                                 [in]
!   rho-- densities of particles                                                       [in]
!   p-- pressure  of particles                                                         [in]
!   pressure_force -- pressure force on each particle                                  [in]
!   viscous_force -- viscous force on each particle                                    [in]
!   ntotal-- total particle number                                                     [in]
!---------------------------------------------------------------------------------------------

     implicit none
     include 'param.inc'


     integer ntotal, aux, i, funit, funit2, funit3, funit4, funit5, funit6, funit7, funit8,
itimestep, recording_step
    double precision x(maxn,2), v(maxn,2), rho(maxn), p(maxn), accel(maxn,2),
pressure_force(maxn,2), viscous_force(maxn,2)
    character(len=25) aux2
    character(len=45) name_pos, name_vel, name_density, name_pres, name_gradient,
name_visc, name_accel
    aux=itimestep
    if ((itimestep.eq.1).or.(mod(itimestep,recording_step).eq.0)) then
```

```
    write(aux2,*) itimestep         ! Converting an integer value to a character

    if (itimestep.eq.1) then
     funit=itimestep+3100
     funit2=itimestep+60100
     funit3=itimestep+91000
     funit4=itimestep+121000
     funit5=itimestep+151000
     funit6=itimestep+181000
     funit7=itimestep+211000
    else
     funit=int(itimestep/recording_step)+3100
     funit2=int(itimestep/recording_step)+60100
     funit3=int(itimestep/recording_step)+91000
     funit4=int(itimestep/recording_step)+121000
     funit5=int(itimestep/recording_step)+151000
     funit6=int(itimestep/recording_step)+181000
     funit7=int(itimestep/recording_step)+211000
    end if

  !-----------------------------------------------------------------------------------------------

    if (aux.lt.10)  name_pos='../output/data/position/000000'//trim(adjustl(aux2))//'.dat'
    if                                           ((aux.ge.10).and.(aux.lt.100))
name_pos='../output/data/position/00000'//trim(adjustl(aux2))//'.dat'
    if                                           ((aux.ge.100).and.(aux.lt.1000))
name_pos='../output/data/position/0000'//trim(adjustl(aux2))//'.dat'
    if                                           ((aux.ge.1000).and.(aux.lt.10000))
name_pos='../output/data/position/000'//trim(adjustl(aux2))//'.dat'
    if                                           ((aux.ge.10000).and.(aux.lt.100000))
name_pos='../output/data/position/00'//trim(adjustl(aux2))//'.dat'
    if                                           ((aux.ge.100000).and.(aux.lt.1000000))
name_pos='../output/data/position/0'//trim(adjustl(aux2))//'.dat'
    if (aux.eq.1000000) name_pos='../output/data/position'//trim(adjustl(aux2))//'.dat'

    open(unit=funit,file=name_pos)
    do i = 1, ntotal
     write (funit,221) i, x(i,1), x(i,2)
    end do

    close (funit)
  !-----------------------------------------------------------------------------------------------

    if (aux.lt.10)  name_vel='../output/data/velocity/000000'//trim(adjustl(aux2))//'.dat'
    if                                           ((aux.ge.10).and.(aux.lt.100))
name_vel='../output/data/velocity/00000'//trim(adjustl(aux2))//'.dat'
    if                                           ((aux.ge.100).and.(aux.lt.1000))
name_vel='../output/data/velocity/0000'//trim(adjustl(aux2))//'.dat'
    if                                           ((aux.ge.1000).and.(aux.lt.10000))
name_vel='../output/data/velocity/000'//trim(adjustl(aux2))//'.dat'
```

```
   if ((aux.ge.10000).and.(aux.lt.100000))
name_vel='../output/data/velocity/00'//trim(adjustl(aux2))//'.dat'
   if ((aux.ge.100000).and.(aux.lt.1000000))
name_vel='../output/data/velocity/0'//trim(adjustl(aux2))//'.dat'
   if (aux.eq.1000000) name_vel='../output/data/velocity'//trim(adjustl(aux2))//'.dat'

   open(unit=funit2,file=name_vel)
   do i = 1, ntotal
     write (funit2,221) i, v(i,1), v(i,2)
   end do

  close (funit2)

!-----------------------------------------------------------------------------------------------

   if (aux.lt.10)  name_density='../output/data/density/000000'//trim(adjustl(aux2))//'.dat'
   if ((aux.ge.10).and.(aux.lt.100))
name_density='../output/data/density/00000'//trim(adjustl(aux2))//'.dat'
   if ((aux.ge.100).and.(aux.lt.1000))
name_density='../output/data/density/0000'//trim(adjustl(aux2))//'.dat'
   if ((aux.ge.1000).and.(aux.lt.10000))
name_density='../output/data/density/000'//trim(adjustl(aux2))//'.dat'
   if ((aux.ge.10000).and.(aux.lt.100000))
name_density='../output/data/density/00'//trim(adjustl(aux2))//'.dat'
   if ((aux.ge.100000).and.(aux.lt.1000000))
name_density='../output/data/density/0'//trim(adjustl(aux2))//'.dat'
   if (aux.eq.1000000) name_density='../output/data/density'//trim(adjustl(aux2))//'.dat'

   open(unit=funit3,file=name_density)
   do i= 1, ntotal
     write (funit3,222) i, rho(i)
   end do

   close (funit3)

!-----------------------------------------------------------------------------------------------

   if (aux.lt.10)  name_pres='../output/data/pressure/000000'//trim(adjustl(aux2))//'.dat'
   if ((aux.ge.10).and.(aux.lt.100))
name_pres='../output/data/pressure/00000'//trim(adjustl(aux2))//'.dat'
   if ((aux.ge.100).and.(aux.lt.1000))
name_pres='../output/data/pressure/0000'//trim(adjustl(aux2))//'.dat'
   if ((aux.ge.1000).and.(aux.lt.10000))
name_pres='../output/data/pressure/000'//trim(adjustl(aux2))//'.dat'
   if ((aux.ge.10000).and.(aux.lt.100000))
name_pres='../output/data/pressure/00'//trim(adjustl(aux2))//'.dat'
   if ((aux.ge.100000).and.(aux.lt.1000000))
name_pres='../output/data/pressure/0'//trim(adjustl(aux2))//'.dat'
   if (aux.eq.1000000) name_pres='../output/data/pressure'//trim(adjustl(aux2))//'.dat'
```

```
    open(unit=funit4,file=name_pres)
    do i = 1, ntotal
     write (funit4,222) i, p(i)
    end do

    close (funit4)

!------------------------------------------------------------------------------------------------------

    if                                                                    (aux.lt.10)
name_gradient='../output/data/pressure_force/000000'//trim(adjustl(aux2))//'.dat'
    if                                                    ((aux.ge.10).and.(aux.lt.100))
name_gradient='../output/data/pressure_force/00000'//trim(adjustl(aux2))//'.dat'
    if                                                  ((aux.ge.100).and.(aux.lt.1000))
name_gradient='../output/data/pressure_force/0000'//trim(adjustl(aux2))//'.dat'
    if                                                ((aux.ge.1000).and.(aux.lt.10000))
name_gradient='../output/data/pressure_force/000'//trim(adjustl(aux2))//'.dat'
    if                                              ((aux.ge.10000).and.(aux.lt.100000))
name_gradient='../output/data/pressure_force/00'//trim(adjustl(aux2))//'.dat'
    if                                            ((aux.ge.100000).and.(aux.lt.1000000))
name_gradient='../output/data/pressure_force/0'//trim(adjustl(aux2))//'.dat'
    if                                                               (aux.eq.1000000)
name_gradient='../output/data/pressure_force'//trim(adjustl(aux2))//'.dat'

    open(unit=funit5,file=name_gradient)
    do i = 1, ntotal
     write (funit5,221) i, pressure_force(i,1), pressure_force(i,2)
    end do

    close (funit5)

!------------------------------------------------------------------------------------------------------

    if                                                                    (aux.lt.10)
name_visc='../output/data/viscous_force/000000'//trim(adjustl(aux2))//'.dat'
    if                                                    ((aux.ge.10).and.(aux.lt.100))
name_visc='../output/data/viscous_force/00000'//trim(adjustl(aux2))//'.dat'
    if                                                  ((aux.ge.100).and.(aux.lt.1000))
name_visc='../output/data/viscous_force/0000'//trim(adjustl(aux2))//'.dat'
    if                                                ((aux.ge.1000).and.(aux.lt.10000))
name_visc='../output/data/viscous_force/000'//trim(adjustl(aux2))//'.dat'
    if                                              ((aux.ge.10000).and.(aux.le.100000))
name_visc='../output/data/viscous_force/00'//trim(adjustl(aux2))//'.dat'
    if                                            ((aux.ge.100000).and.(aux.lt.1000000))
name_visc='../output/data/viscous_force/0'//trim(adjustl(aux2))//'.dat'
    if                                                               (aux.eq.1000000)
name_visc='../output/data/viscous_force'//trim(adjustl(aux2))//'.dat'

    open(unit=funit6,file=name_visc)
```

```
    do i = 1, ntotal
      write (funit6,221) i, viscous_force(i,1), viscous_force(i,2)
    end do

    close (funit6)
!------------------------------------------------------------------------------------

    if                                                                    (aux.lt.10)
name_accel='../output/data/acceleration/000000'//trim(adjustl(aux2))//'.dat'
    if                                                       ((aux.ge.10).and.(aux.lt.100))
name_accel='../output/data/acceleration/00000'//trim(adjustl(aux2))//'.dat'
    if                                                     ((aux.ge.100).and.(aux.lt.1000))
name_accel='../output/data/acceleration/0000'//trim(adjustl(aux2))//'.dat'
    if                                                   ((aux.ge.1000).and.(aux.lt.10000))
name_accel='../output/data/acceleration/000'//trim(adjustl(aux2))//'.dat'
    if                                                 ((aux.ge.10000).AND.(aux.lt.100000))
name_accel='../output/data/acceleration/00'//trim(adjustl(aux2))//'.dat'
    if                                               ((aux.ge.100000).and.(aux.lt.1000000))
name_accel='../output/data/acceleration/0'//trim(adjustl(aux2))//'.dat'
    if                                                                (aux.eq.1000000)
name_accel='../output/data/acceleration'//trim(adjustl(aux2))//'.dat'

    open(unit=funit7,file=name_accel)
    do i= 1, ntotal
      write (funit7,221) i, accel(i,1), accel(i,2)
    end do

    close (funit7)

    221   format (I8, 2(1x,D21.14))
    222   format (I8,1x,D21.14)

  end if

 end subroutine output


!------------------------------------------------------------------------------------
! Here are the parameters and constants used in the SPH software
! A file must be created with the name param.inc
!------------------------------------------------------------------------------------


!   maxn    : Maximum number of particles
    integer maxn, n_vizinhas
    parameter ( maxn  = 100000, n_vizinhas = 50 )

!    alpha coefficient (in artificial viscosity calculation)
     double precision
     parameter ( alpha = 0.20 )
```

```
!     XSPH Correction
      character (len = 3) XSPH_correction
      parameter (XSPH_correction = 'NO')

      double precision pi
      parameter ( pi = 3.14159265358979323846 )


#
=============================================================
========
# makefile
#
=============================================================
========
/folder_address_of_item_2/fonts/main.run: geometry.o input.o direct_find.o kernel.o
continuity_density.o pressure. o internal_force.o \
external_force.o acceleration.o XSPH.o time_integration.o collision_test.o output.o \

#
        ifort -O3 -o /folder_address_of_item_2/fonts/main.run geometry.o input.o
direct_find.o kernel.o continuity_density.o \
pressure.o internal_force.o external_force.o acceleration.o XSPH.o time_integration.o
collision_test.o output.o \

#
#

main.o:         param.inc  main.f90
        ifort -O3 -c  main.f90
geometry.o:
        ifort -O3 -c  geometry.f90
input.o:  param.inc input.f90
        ifort -O3 -c  input.f90
direct_find.o:
        ifort -O3 -c  direct_find.f90
kernel.o:  param.inc kernel.f90
        ifort -O3 -c  kernel.f90
continuity_density.o:  param.inc  continuity_density.f90
        ifort -O3 -c  continuity_density.f90
pressure.o:  param.inc pressure.f90
        ifort -O3 -c  pressure.f90
internal_force.o:  param.inc internal_force.f90
        ifort -O3 -c  internal_force.f90
external_force.o:
        ifort -O3 -c  external_force.f90
acceleration.o:  param.inc acceleration.f90
        ifort -O3 -c  acceleration.f90
XSPH.o:  param.inc XSPH.f90
        ifort -O3 -c  XSPH.f90
```

```
time_integration.o:  param.inc time_integration.f90
        ifort -O3 -c  time_integration.f90
collision_test.o:
        ifort -O3 -c  collision_test.f90
output.o:
        ifort -O3 -c  output.f90
#
=====================================================================
# clean
#
=====================================================================
clean:
        rm -rf *.o
```

Guidelines for implementing the program:

1. Create individual files (*name.f90*) for the *main program* and *subroutines*, besides *param.inc* and *Makefile*
2. Create a folder with the chosen name to paste all files
3. Inside the previous folder, create the *fonts*, *output* and *figures* folders
4. Inside *output*, create the *data* folder with the folders named *acceleration*, *density*, *position*, *pressure*, *pressure_force*, *velocity* and *viscous_force*
5. Create a file named *Makefile* with the content presented above
6. Set the folder address of item 2 in the *Makefile*
7. Open the terminal inside */folder_address_of item_2/fonts/* and give the command make
8. Run the program with the command ./main.run

Note: Inside the output folder there is a script to generate graphs showing the evolution of fluid particles over time.

Chapter 4
Three-Dimensional Dam-Breaking over a Dry Bed

```
program main

!  A file named "main.f90" can be created

!---------------------------------------------------------------------------------------------------
!  Program for 3D simulation od dam-breaking using the SPH method coupled
!  with reflective boundary conditions.
!
! The following are the basic parameters needed for the operations in this code or
!  calculated  by this code
!  mass -- mass of particles                                          [in]
!  ntotal -- total particle number                                    [in]
!  dt -- time step used in the time integration                       [in]
!  x -- coordinates of particles                                  [in/out]
!  v -- velocities of particles                                   [in/out]
!  rho -- densities of particles                                  [in/out]
!  p -- pressure  of particles                                    [in/out]
!  hsml --  smoothing lengths of particles                        [in/out]
!  c_sound -- sound velocity of particles                            [out]
!---------------------------------------------------------------------------------------------------

   implicit none
   include 'param.inc'
   integer itype(maxn), neighbours(maxn,n_neighbours), itimestep, maxtimestep,ntotal,
d, m, i, k, hours, minutes, hours_in_seconds, nx, ny, nz, mp, np, wp, recording_step,
aux, surface_particles(maxn)
    double precision x(maxn,dim), v(maxn,dim),
gravity(maxn,dim),mass(maxn),rho(maxn), p(maxn), hsml(maxn), viscosity(maxn),
r(maxn,n_neighbours), dx(maxn,n_neighbours), dy(maxn,n_neighbours),
dz(maxn,n_neighbours), w(maxn,n_neighbours), dwdx(maxn,n_neighbours),
dwdy(maxn,n_neighbours), dwdz(maxn, n_neighbours), dwdr(maxn, n_neighbours), dt,
factor, sum_CSPM(maxn), viscous_force(maxn,dim), pressure_force(maxn,dim),
accel(maxn,dim), v_xsph(maxn,dim), drhodt(maxn)
    double precision start, finish, time, seconds, radius, xl, yl, zl, xlmin, xlmax, ylmin,
ylmax,zlmin, zlmax, dx1, dy1, dz1, kh, K_scale, vmax, grav, Mach, b, c_sound, CR,
friction_coef
    double precision Fij_art_pressure(maxn,n_neighbours), Rp_ij(maxn,n_neighbours)
  !-----------------------------------------------------------------------
  !------------------- Simulated geometry --------------------------
  !-----------------------------------------------------------------------
   ! --- Reservoir coordinates
    xlmin=0.
    ylmin=0.
```

Supplementary Information The online version contains supplementary material available at https://doi.org/10.1007/978-3-031-99370-1_4.

C. A. Dutra Fraga Filho, *Computational Algorithms for Reflective Boundary Conditions in SPH Fluid Dynamics Simulation*, Synthesis Lectures on Mechanical Engineering, https://doi.org/10.1007/978-3-031-99370-1_4

```
  zlmin=0.
  xlmax=0.42
  ylmax=0.44
  zlmax=0.228
  grav=9.81
  Mach= 0.10

  !--------Fluid inside the tank (initial configuration)----
  xl=0.114
  yl=2*xl
  zl = zlmax
! -------- Initial conditions
  vmax= sqrt(2*grav*yl)
  c_sound=vmax/Mach

  m=5          ! number of particles in each side of the cell
  nx=4         ! number of cells in x-direction
  ny=nx*int(yl/xl)  ! number of cells in y-direction
  nz=nx*int(zl/xl)  ! number of cells in z-direction

  mp=m*nx                  ! number of particles in each the x-direction
  np=m*ny                  ! number of particles in each the y-direction
  wp=m*nz                  ! number of particles in each the z-direction
  ntotal = np*mp*wp  ! total number of particles in the domain
  CR = 1.0              ! coefficient of restitution of kinetic energy
  friction_coef = 0.   ! coefficient of friction
  dt = 1.e-4            ! simulation time step
  maxtimestep = 10000   ! mximum number of simulation steps
  recording_step = 10   ! step to record output files
  print*,'------------------------------------------------------------------------------'
  print*, ' ------------------------- INITIAL CONDITIONS -------------------------'
  print*, '------------------------------------------------------------------------------'
  print*, ' TANK LENGTH - x (m) = ', xlmax
  print*, ''
  print*, ' TANK HEIGHT - y (m) = ', ylmax
  print*, ''
  print*, ' TANK DEPTH - z(m) = ', zlmax
  print*, ''
  print*, ' LENGTH OF WATER IN THE TANK - x(m) = ', xl
  print*, ''
  print*, ' WATER COLUMN HEIGHT - h (m) = ' , yl
  print*, ''
  print*, ' WATER DEPTH - z (m) = ' , zl
  print*, ''
  print*,'The number of water particles used in this simulation = ', ntotal
  print*, '-----------------------------------------------------------------------'
  print*, ''
  pause

  open(unit=997, file='../output/largest_x.dat')
  open(unit=998, file='../output/geometry.dat')
```

```
    open(unit=999, file='../output/initial_distribution.dat')
    open(unit=1000, file='../output/timestep.dat')

    write(998,554) xlmin, xlmax
    write(998,554) ylmin, ylmax
    write(998,554) zlmin, zlmax
554 format(1x,4(E12.4))

    call cpu_time(start)

    open(unit=445,file='../output/normals.dat')
    open(unit=450,file='../output/flat_points.dat')

    open(40,file="../output/initial_velocity.dat")
    open(41,file="../output/initial_position.dat")
    open(45,file="../output/velocity.dat")
    open(46,file="../output/position.dat")
    open(74,file='../output/initial_properties.dat')
    open(75,file='../output/properties.dat')

    rewind(998)
    rewind(999)
    rewind(1000)
    write(999,*) ntotal
    write(1000,*) dt
    write(1000,*) recording_step
    write(1000,*) maxtimestep
    close(998)
    close(999)
    close(1000)

    aux=1
    itimestep = aux-1

    call geometry(xlmax,ylmax,zlmax)

    call input (xl, yl, zl, radius, xlmin, xlmax, ylmin, ylmax, x,m, mp, np, wp, ntotal, nx,
dx1, dy1, dz1, kh, mass, rho, itype, hsml, K_scale, p, viscosity, surface_particles)

    do i=aux,maxtimestep

      itimestep=itimestep+1

    !----------------- CREATING FILE WITH DATA FOR POST-PROCESSING ---------

      if (itimestep.eq.1) then
        open(17,file='../output/matlab_parameters.dat')
        rewind(17)
        write(17,18) xlmin, xlmax, ylmin, ylmax, zlmin, zlmax
        write(17,19) ntotal, 0, 0, 0, 0, 0
```

```
18        format (6(1x,e10.3))
19        format (6(1x,I8))
        close(17)
      end if

      if (itimestep.ne.1) call h_upgrade(dt,ntotal, rho, drhodt, hsml, viscosity)

      call direct_find(itimestep, ntotal, hsml,neighbours,x, r,dx,dy,dz,kh, K_scale)

      call kernel(itimestep,itype, ntotal, r,hsml,mass, rho, w, neighbours,dx, dy, dz,
dwdx,dwdy,dwdz, dwdr, sum_CSPM,factor,Fij_art_pressure, dx1)

      call continuity_density(itimestep,ntotal, mass,dwdx,dwdy,dwdz, v,itype,x,rho, r,
hsml, dx, dy, dx1, dy1, dz1, neighbours, w, dt, drhodt, recording_step)

      call pressure(itimestep, ntotal, x, rho, c_sound, mass, viscosity, p, w, itype,
neighbours, surface_particles, K_scale, hsml, Rp_ij)

      call internal_force(itimestep,ntotal,hsml,mass,v,neighbours,r,factor, p, c_sound,
pressure_force, viscous_force,rho,viscosity,dx,dy,dz, dwdx,dwdy, dwdz,
dwdr,recording_step, Rp_ij, Fij_art_pressure)

      call external_force(ntotal,gravity)

      call acceleration(itimestep, mass, ntotal, pressure_force, viscous_force, gravity,
accel, recording_step)
      if (XSPH_correction.eq."YES") then ! The choice to apply XSPH correction  (see
param.inc)
        call XSPH(ntotal,mass,neighbours, w, rho, v , v_xsph)
      end if

      call time_integration(itimestep,accel,ntotal,dt,x,v,v_xsph,recording_step)

      call free_surface(itimestep,ntotal,recording_step,surface_particles,p,x,itype)

      call collision_test(ntotal,radius,itimestep,dt,CR,friction_coef)

      call output(ntotal,itimestep,recording_step,p,hsml)
      print*, ' '
      print*, '-----------------------------------------------------------------------'
      print*, ' LEAVING THE ITERATION NUMBER ', itimestep, ' OUT OF A
TOTAL OF ', maxtimestep
      print*, '-----------------------------------------------------------------------'
      print*, ' '
      print*, ' Simulated physical time (s) = ', i*dt
      print*, ' '

    end do
    close(45)
    close(46)
```

```
    close(445)
    close(450)
    close(75)
    close(1001)
    close(881)

    call cpu_time(finish)
    time=finish-start
    hours_in_seconds=3600
    hours = (time/hours_in_seconds)
    minutes = (time -(hours_in_seconds*hours))/60
    seconds = (time -(hours_in_seconds*hours)-(minutes*60))
    open(unit=807, file='../output/SIMULATION_PARAMETERS.dat')
    write(807,*) ' ---------------------------------------------------------------------'
    write(807,*) ''
    write(807,*) ' TANK LENGTH  (m) = ', xlmax
    write(807,*) ''
    write(807,*) ' TANK HEIGHT - y (m) = ', ylmax
    write(807,*) ''
    write(807,*) ' TANK DEPTH - z(m) = ', zlmax
    write(807,*) ''
    write(807,*) ' WIDTH OF WATER IN TANK (m) = ', xl
    write(807,*) ''
    write(807,*) ' WATER HEIGHT IN TANK (m) = ' , yl
    write(807,*) ''
    write(807,*) ' WATER DEPHT IN TANK (m) = ' , zl
    write(807,*) ''
    write(807,*) ' Total number of water particles in the simulation :  '
    write(807,*) mp, 'X', np ,'X',  wp, ' = ', np*mp*wp
    write(807,*) ' '
    write(807,*) 'Timestep (s) = ', dt
    write(807,*)''
    write(807,*) 'Distance between two consecutive particles centres of mass  (m) = ',
dx1
    write(807,*)''
    write(807,*) 'Simulated Physical Time (s) =', i*dt
    write(807,*)'  '
    write(807,*)''
    write(807,*) 'alpha (coefficient in artificial viscosity) =', alpha
    write(807,*)'  '
    write(807,*) 'CPU time  = '
    write(807,*) hours ,' hours', minutes,' minutes', seconds, ' seconds.'
    write(807,*) ' ---------------------------------------------------------------------'

    print*, ' ---------------------------------------------------------------------'
    print*, 'END OF SIMULATION'
    print*, ''
    print*, ' TANK LENGTH  (m) = ', xlmax
    print*, ''
    print*, ' TANK HEIGHT - y (m) = ', ylmax
```

```
    print*, ''
    print*, ' TANK DEPTH - z(m) = ', zlmax
    print*, ''
    print*, ' WIDTH OF WATER IN TANK (m) = ', xl
    print*, ''
    print*, ' WATER HEIGHT IN TANK (m) = ' , yl
    print*, ''
    print*, ' WATER DEPHT IN TANK (m) = ' , zl
    print*, ''
    print*, 'Total number of water particles in the simulation : '
    print*, mp, 'x', np ,'x',  wp, ' = ', np*mp*wp
    print*, ''
    print*, 'Timestep (s) =', dt
    print*,''
    print*, 'Simulated Physical Time (s) = ', i*dt
    print*,'           '
    print*, 'CPU time  = '
    write(*,*) hours ,' hours', minutes,' minutes', seconds, ' seconds.'
    print*, ' -----------------------------------------------------------------'

end program main

!---------------------------------------------------------------------------------------------
! Subroutines:
! Files should be created with the name "subroutine_name.f90", for example,
! geometry.f90
!---------------------------------------------------------------------------------------------

subroutine geometry(xlmax,ylmax,zlmax)

!---------------------------------------------------------------------------------------------
!  In this subroutine, the planes that make up the flow boundaries are defined.
!---------------------------------------------------------------------------------------------
!            (y)
!              I
!             I
!            I
!           I I I I I (x)
!       I
!     I
!   I (z)
!

    implicit none
    include 'param.inc'

    integer i,j,active_planes(6)
    double precision xlmax, ylmax,zlmax,n(6,dim),X_0(6,dim),summation,aux
```

```
!--- DEFINING NORMAL VECTORS TO PLANES
!--- NOTE: IF YOU WANT TO DISABLE A PLANE, JUST DEFINE THE NORMAL
VECTOR TO THE PLANE AS n_0=(0.,0.,0.
rewind(445)
rewind(450)

!----------------------------------------------------------------------------------------------------
!------------------------PLANES THAT DEFINE THE RESERVOIR ----------------------
!----------------------------------------------------------------------------------------------------

! The normal vector to the right lateral plane is (-1.,0.,0.) - **plane 01** - perpendicular
to the zy plane!
   n(1,1)=-1.
   n(1,2)=0.
   n(1,3)=0.

! The normal vector to the left lateral plane is (1.,0.,0.) - **plane 02** - perpendicular to
the zy direction!
   n(2,1)=1.
   n(2,2)=0.
   n(2,3)=0.

! The normal vector to the lower plane is (0.,1.,0.) - **plane 03** - perpendicular to the
zx plane!
   n(3,1)=0.
   n(3,2)=1.
   n(3,3)=0.

! The normal vector to the upper plane is (0.,0.,0.) - **plane 03** - perpendicular to the
zx plane ! ----- DISABLED --------
   n(4,1)=0.
   n(4,2)=0.
   n(4,3)=0.

! The normal vector to the back plane (lower z) is (0.,0.,1.) - **plane 05** - perpendicular
to the yx direction!
   n(5,1)=0.
   n(5,2)=0.
   n(5,3)=1.

! The normal vector to the frontal plane (larger z) is (0.,0.,-1.) - **plane 06** -
perpendicular to the yx direction!
   n(6,1)=0.
   n(6,2)=0.
   n(6,3)=-1.

!----------------------------------------------------------------------------------------------------
! DEFINING A POINT BELONGING TO EACH PLANE !
! ----- ATTENTION: HERE YOU DEFINE THE GEOMETRY OF THE RESERVOIR!
!----------------------------------------------------------------------------------------------------
```

```
! X_0=(xlmax,0.,0.) belongs to the right lateral plane
   X_0(1,1)=xlmax
   X_0(1,2)=0.
   X_0(1,3)=0.

! X_0=(0.,0.,0.) belongs to the left lateral plane
   X_0(2,1)=0.0
   X_0(2,2)=0.0
   X_0(2,3)=0.0

! X_0=(0.,0.,0.) belongs to the lower plane
   X_0(3,1)=0.0
   X_0(3,2)=0.0
   X_0(3,3)=0.0

! X_0=(0.,ylmax,0.) belongs to the upper plane
   X_0(4,1)=0.
   X_0(4,2)=ylmax
   X_0(4,3)=0.

! X_0=(0.,0.,0.) belongs to the back plane
   X_0(5,1)=0.
   X_0(5,2)=0.
   X_0(5,3)=0.

! X_0=(0,0,zlmax) belongs to the frontal plane
   X_0(6,1)= 0.
   X_0(6,2)=0.
   X_0(6,3)=zlmax

   do i=1,6
     write(445,13) (n(i,j),j=1,dim)
     write(450,13) (X_0(i,j),j=1,dim)
13 format(3(2x,F10.3))
end do

  do i=1,6
    summation=0.
   do j=1,dim
     summation = summation + abs(n(i,j))
   end do
   active_planes(i)=summation     ! If plane i is active, the value of its position in the
vector will be different from zero
  end do

! ---------- Checking active planes ------------
  print*, '---------------------------------------------------------------------------'
  print*, '-------------------------- ACTIVE PLANES -----------------------------'
  print*, '---------------------------------------------------------------------------'
```

```
  do i=1,6

   if (active_planes(i).ne.0.) then

    if (i.eq.1) then
     print*, '[1] - [RESERVOIR: RIGHT SIDE]'
    else
     if (i.eq.2) then
      print*, '[2] - [RESERVOIR: LEFT SIDE]'
     else
      if (i.eq.3) then
       print*, '[3] - [RESERVOIR: BOTTOM]'
      else
       if (i.eq.4) then
        print*, '[4] - [RESERVOIR: UPPER]'
       else
        if (i.eq.5) then
         print*, '[5] - [RESERVOIR: BACK PLANE]'
        else
         if (i.eq.6) then
          print*, '[6] - [RESERVOIR: FRONTAL PLANE]'
         end if
        end if
       end if
      end if
     end if
    end if

   end if

  end do

end subroutine geometry


subroutine input (xl, yl, zl, radius, xlmin, xlmax, ylmin, ylmax, x,m, mp, np, wp, ntotal,
nx, dx1, dy1, dz1, kh, mass, rho, itype, hsml, K_scale, p, viscosity, surface_particles)

!----------------------------------------------------------------------------------------------------
!  Subroutine for generating initial particle distribution
!  x -- coordinates of particles                                              [out]
!  mass-- mass of particles                                                   [out]
!  rho-- densities of particles                                               [out]
!  hsml-- smoothing lengths of particles                                      [out]
!  ntotal-- total domain particle number                                      [out]
!----------------------------------------------------------------------------------------------------
   implicit none
   include 'param.inc'
```

```
   double precision  grav, xl, yl, zl, yl_oleo, xlmin, xlmax, ylmin, ylmax, dx1, dy1, dz1,
kh, K_scale, radius
   integer i, j, k,t, d, nx, ny, m,n,mp, np, wp, ntotal, surface_particles(maxn), cont, aux
   integer  itype(maxn)
   double precision  mass(maxn), rho(maxn), hsml(maxn), p(maxn), viscosity(maxn),
max_x, max_y
   double precision  x(maxn,dim), v(maxn,dim)

   grav=9.81
   dx1 = real((xl/nx)/m)
   dy1 = dx1
   dz1 = dy1
   radius = dx1/2.
   kh= real((xl/nx)/2.)
   K_scale = 2.   ! Normalization constant in the cubic spline kernel

   rewind(40)
   rewind(41)
   rewind(45)
   rewind(46)
   rewind(74)
   rewind(75)

   !--------------- Domain Particle Distribution and Initial Properties -----------------

   open(876, file='../output/data/position/0000000.dat')
   open(877, file='../output/data/velocity/0000000.dat')
   open(878, file='../output/data/pressure/0000000.dat')
   open(879, file='../output/data/free_surface/0000000.dat')

   do i=1,ntotal
     surface_particles(i)=0
   end do

   k=0
   do t = 1, wp
     do i = 1, mp
       do j = 1, np
         k=k+1
         x(k,3) =  (t-1)*dz1 + dz1/2.
         x(k,2) =  (j-1)*dy1 + dy1/2.
         x(k,1) =  (i-1)*dx1 + dx1/2.

         if (x(k,1).gt.max_x)   max_x = x(k,1)
           if (x(k,2).gt.max_y)   max_y = x(k,2)
             v(k,1) = 0.
             v(k,2) = 0.
             v(k,3) = 0.
             rho (k) = 1000.
             mass(k) = dx1*dy1*dz1*rho(k)
```

```
            viscosity(k)=1.e-3
            hsml(k) = kh/K_scale
            itype(k)=2
            write(876,222) k, x(k,1), x(k,2),  x(k,3)
            write(877,222) k, v(k,1), v(k,2),  v(k,3)

        end do
      end do
    end do

   do i=1, k

     if ( (x(i,1).eq.max_x).OR.(x(i,2).eq.max_y) )  then
       p(i)=0.
       surface_particles(i)=i
       write (879,225) i, x(i,1), x(i,2), x(i,3)
225    format (I8, 3(1x,D21.14))
     end if
   end do

   k=0
   do t = 1, wp
     do i = 1, mp
       do j = 1,np
         k=k+1
         p(k)= rho(k) * grav * (yl-x(k,2))
         write(878,55)  k, rho (k), mass(k), p(k)
       end do
     end do
   end do
222    format (I6, 3(1x,D21.14))

   close(881)
   close(876)
   close(877)
   close(878)
   close(879)

   do k=1,ntotal
     write(41,54) k, (x(k,d), d = 1,3)
     write(40,54) k, (v(k,d), d = 1,3)
     write(46,54) k, (x(k,d), d = 1,3)
     write(45,54) k, (v(k,d), d = 1,3)
     write(74,56) k, x(k,1), x(k,2), x(k,3), rho (k), mass(k), p(k)
     write(75,55) k, rho (k), mass(k), p(k)
   end do

54  format(I8, 3(1x,D21.14))
55  format(I8,3(1x,D21.14))
56  format(I8, 6(1x,e11.4))
```

```
  write(*,*)' '
  write(*,*)'     Total number of particles  = ', ntotal
  write(*,*)' '
  close(40)
  close(41)
  close(74)

  open(unit=880,file='../output/data/free_surface.dat')
  do i=1,ntotal
    write(880,543) surface_particles(i)
  end do
 543 format (I8)
  close (880)

end subroutine input


subroutine  h_upgrade(dt,ntotal, rho, drhodt, hsml, viscosity)

!------------------------------------------------------------------------------------------
!  Subroutine for updating the smoothing radius
!  dt -- time step                                                              [in]
!   ntotal -- number of particles                                               [in]
!   mass  -- particle masses                                                    [in]
!   rho -- density                                                              [in]
!   niac -- number of interaction pairs                                         [in]
!   hsml -- smoothing length                                                [in/out]
!   drhodt -- density change rate of each particle                              [in]
!------------------------------------------ ----------------------------------------------

    implicit none
    include 'param.inc'

    integer ntotal
    double   precision   hsml(maxn),   rho(maxn),      viscosity(maxn),   drhodt(maxn),
dhsmldt(maxn)
    integer i,j,k,d
    double precision dt

    do i=1, ntotal
      dhsmldt(i) = (-1/dim)*(hsml(i)/rho(i))*drhodt(i)
    end do

    ! Integrating and recording the smoothing lenght
    do k=1,ntotal
      hsml(k)=hsml(k) + dhsmldt(k)*dt
      if (hsml(k).le.0) hsml(k) = hsml(k) - dt*dhsmldt(k)
    end do
```

```
end subroutine h_upgrade


subroutine direct_find(itimestep, ntotal, hsml,neighbours,x, r,dx,dy,dz,kh, K_scale)

!---------------------------------------------------------------------------------------------------
!   Subroutine to detect and store neighboring particles.
!   Interaction pairs are determined by directly comparing the particles distance with
!   the corresponding support radius.
!
!   ntotal --  number of particles in the domain                                   [in]
!   hsml   --  smoothing length                                                     [in]
!   x   --  coordinates of all particles                                            [in]
!  dx -- x-distance between particles i and j                                       [in]
!  dy -- y-distance between particles i and j                                       [in]
!  niac  --  number of neighbours                                                  [out]
!  distance --  distance between two particles                                     [out]
!  neighbours -- matrix with the neighbors of each particle in the domain          [out]
!---------------------------------------------------------------------------------------------------

   implicit none
   include 'param.inc'

   integer  i, j, k, itimestep, ntotal, niac, neighbours(maxn,n_neighbours)
   double precision aux_distance, kh, K_scale
   double  precision    x(maxn,dim),  hsml(maxn),  r(maxn,n_neighbours),  distance,
dx(maxn,n_neighbours), dy(maxn,n_neighbours), dz(maxn,n_neighbours)

   do i=1,ntotal
    do j=1,n_neighbours
     neighbours(i,j)=0
     dx(i,j)=0.
     dy(i,j)=0.
     dz(i,j)=0
     r(i,j)=0.
    end do
   end do

   do i=1,ntotal
    niac= 2
    neighbours(i,1)=i

    do j = 1, ntotal
     aux_distance  = (x(j,1) - x(i,1))**2 + (x(j,2) - x(i,2))**2+ (x(j,3) - x(i,3))**2
     distance = sqrt(aux_distance)

     if (distance.le.(hsml(i)*K_scale) ) then
       neighbours(i,niac)=j
       r(i,niac)=distance
       dx(i,niac)=x(j,1) - x(i,1)
```

```
        dy(i,niac)=x(j,2) - x(i,2)
        dz(i,niac)=x(j,3) - x(i,3)
        niac=niac+1
      end if

    end do

  end do

  open(222,file='../output/neighbours.dat')  ! Storing the neighbors of each particle in
the domain in the current iteration
  do i=1,ntotal
    write(222,*) (neighbours(i,j), j=1,n_neighbours)
  end do
  close(222)

end subroutine direct_find


subroutine  kernel(itimestep,itype,  ntotal,  r,hsml,mass,  rho,  w,  neighbours,dx,  dy,  dz,
dwdx,dwdy,dwdz, dwdr, sum_CSPM,factor,Fij_art_pressure, dx1)

!----------------------------------------------------------------------------------------------------
!  Subroutine to calculate the smoothing function (kernel) and its derivatives
!
!  if skf = 1, Cubic spline kernel (defined in param.inc)
!          = 2, Quintic kernel
!          = 3, Quartic kernel by Liu and Lam
!  ntotal --  number of particles in the domain                                          [in]
!  r   -- distance between particles i and j                                             [in]
!  dx -- x-distance between particles i and j                                            [in]
!  dy -- y-distance between particles i and j                                            [in]
!  hsml -- smoothing length                                                              [in]
!  w – kernel value for all interaction pairs                                            [out]
!  factor -- kernel's normalisation constant                                             [out]
!  dwdr -- derivative of kernel in polar coordinates                                     [out]
!  dwdx -- derivative of kernel with respect to x                                        [out]
!  dwdy -- derivative of kernel with respect to y                                        [out]
!  dwdz -- derivative of kernel with respect to z                                        [out]
!  Fij_art_pressure -- force related to the artificial pressure                          [out]
!----------------------------------------------------------------------------------------------------

  implicit none
  include 'param.inc'

  integer i, j, k, niac, ntotal, itype(maxn), neighbours(maxn,n_neighbours), itimestep
  double precision r(maxn,n_neighbours), w(maxn,n_neighbours),
dx(maxn,n_neighbours), dy(maxn,n_neighbours), dz(maxn,n_neighbours), hsml(maxn),
dwdx(maxn,n_neighbours),  dwdy(maxn,n_neighbours),  dwdz(maxn,n_neighbours),
```

```
dwdr(maxn,n_neighbours),      mass(maxn),      rho(maxn),      sum_CSPM(maxn),
w_art_pressure(maxn,n_neighbours), Fij_art_pressure(maxn,n_neighbours)
   double precision q, q_art_pressure, factor, aux ,dx1

   open(88,file='../output/kernel_values.dat')
   rewind(88)

   do i=1, ntotal
    niac=2
    w(i,1)=0.
    dwdx(i,1)=0.
    dwdy(i,1)=0.
    dwdz(i,1)=0.
    sum_CSPM(i)=0.       ! CSPM CORRECTION

    do while (neighbours(i,niac).ne.0)
     j=neighbours(i,niac)

     q = r(i,niac)/hsml(i)
     q_art_pressure = dx1/hsml(i)

     if (dim.eq.1) then           ! Cubic spline kernel
      factor = 1./hsml(i)
     else
      if (dim.eq.2) then
       factor = 15./(7.*pi*hsml(i)*hsml(i))
      else
       if (dim.eq.3) then
        factor = 3./(2.*pi*hsml(i)*hsml(i)*hsml(i))
       end if
      end if
     end if

     if ( (q.ge.0.).and.(q.le.1.) ) then

      w(i,niac)= factor* ((2./3.) - q**2. + (1./2.)*q**3)

     !---------------------------------------------------------------------------------------
     ! FOR CALCULATION OF ARTIFICIAL PRESSURE:
     !---------------------------------------------------------------------------------------
      if (Artificial_press.eq.'YES') then
       q_art_pressure = dx1/hsml(i)
       w_art_pressure(i,niac)   =   factor*   ((2./3.)   -   q_art_pressure**2.   +
(1./2.)*q_art_pressure**3)
       Fij_art_pressure(i,niac) =  w(i,niac)/w_art_pressure(i,niac)
      end if
      !--------------------------------------------------------------------------------

      dwdr(i,niac)      =      factor      *      ((-2.*r(i,niac))/hsml(i)**2      +
(3./2.)*(r(i,niac)**2/(hsml(i)**3)) )
```

```
        dwdx(i,niac) =  factor * ((-2.)/hsml(i)**2 + (3./2.)*(r(i,niac)/(hsml(i)**3)) ) *
dx(i,niac)
        dwdy(i,niac) =  factor * ((-2.)/hsml(i)**2 + (3./2.)*(r(i,niac)/(hsml(i)**3)) ) *
dy(i,niac)
        dwdz(i,niac) =  factor * ((-2.)/hsml(i)**2 + (3./2.)*(r(i,niac)/(hsml(i)**3)) ) *
dz(i,niac)
        sum_CSPM(i)= sum_CSPM(i) + (mass(j)/rho(j))*w(i,niac)

        write(88,12) q, w(i,niac)

       else
        if ((q.gt.1.).and.(q.le.2.)) then
          w(i,niac) = factor * 1.e0/6.e0 * (2.-q)**3

          if (Artificial_press.eq.'YES') then
            q_art_pressure = dx1/hsml(i)
            w_art_pressure(i,niac)   =   factor*   ((2./3.)   -   q_art_pressure**2.   +
(1./2.)*q_art_pressure**3)
            Fij_art_pressure(i,niac) = w(i,niac)/w_art_pressure(i,niac)
          end if

          dwdr(i,niac) = -factor * (1./(2.*hsml(i))) * ( ( 2. - r(i,niac)/hsml(i))**2. )
          dwdx(i,niac) =  -factor * (1./(2.*hsml(i))) * ( ( 2. - r(i,niac)/hsml(i))**2. ) *
dx(i,niac)/r(i,niac)
          dwdy(i,niac) = -factor * (1./(2.*hsml(i))) * ( ( 2. - r(i,niac)/hsml(i))**2. )  *
dy(i,niac)/r(i,niac)
          dwdz(i,niac) = -factor * (1./(2.*hsml(i))) * ( ( 2. - r(i,niac)/hsml(i))**2. )  *
dz(i,niac)/r(i,niac)
          sum_CSPM(i)= sum_CSPM(i) + (mass(j)/rho(j))*w(i,niac)
           write(88,12) q, w(i,niac)
        end if

       end if
       niac=niac+1

     end do

   end do

12 format(2(1x,D21.14))
   close(88)

end subroutine kernel


subroutine continuity_density(itimestep,ntotal, mass,dwdx,dwdy,dwdz, v,itype,x,rho, r,
hsml, dx, dy, dx1, dy1, dz1, neighbours, w, dt, drhodt, recording_step)

!---------------------------------------------------------------------------------------------------------
!  Subroutine to calculate the density with SPH continuity approach.
```

```
!
!   ntotal -- number of particles                                        [in]
!   mass -- particle masses                                              [in]
!   niac  -- number of interaction pairs                                 [in]
!   dwdx -- derivation of Kernel for all interaction pairs               [in]
!   dwdy -- derivative of kernel with respect to y                      [out]
!   dwdz -- derivative of kernel with respect to z                      [out]
!   v  -- velocities of all particles                                    [in]
!   x  -- coordinates of all particles                                   [in]
!   rho -- density                                                       [in]
!   drhodt -- density change rate of each particle                     [out]
!----------------------------------------------------------------------------

    implicit none
    include 'param.inc'

    integer ntotal, niac, itype(maxn), itimestep, aux, funit, recording_step
    double precision mass(maxn), v(maxn,dim), x(maxn,dim), hsml(maxn), rho(maxn),
drhodt(maxn), w(maxn,n_neighbours), r(maxn,n_neighbours), dx(maxn,n_neighbours),
dy(maxn,n_neighbours), dwdx(maxn,n_neighbours), dwdy(maxn,n_neighbours),
dwdz(maxn,n_neighbours)
    integer i,j,k,d, neighbours(maxn,n_neighbours)
    double precision  factor, sum_CSPM(maxn), vcc, dv(dim), dt, rho0, dx1, dy1, dz1,
sum_kernel(maxn)
    character(len= 25 ) aux2
    character(len=45) filename

    do i = 1, ntotal
     drhodt(i) = 0.
    end do

    do k=1,ntotal
     i=neighbours(k,1)
     niac=2

     do while (neighbours(i,niac).ne.0)
       j=neighbours(i,niac)

       do d=1,dim
         dv(d) = v(i,d) - v(j,d)
       end do
       vcc= dv(1)*dwdx(i,niac)
       vcc= vcc+ dv(2)*dwdy(i,niac)
       vcc = vcc + dv(3)*dwdz(i,niac)

       if (itype(i).ne.itype(j)) then
         drhodt(i) = drhodt(i) + (rho(i)/rho(j))*mass(j)*vcc
       else
         drhodt(i) = drhodt(i) + mass(j)*vcc
       end if
```

```
      niac=niac+1

    end do

  end do

  dy1=dx1

!-----------------------------------------------------------------------------------------
!   Integrating and recording the density
!-----------------------------------------------------------------------------------------

  do k=1,ntotal
    rho(k)=rho(k) + drhodt(k)*dt

    if (mod(itimestep,30).eq.0) mass (k) = rho(k)*dx1*dy1*dz1      ! To apply the
Sheppard filter every 30 iterations

  end do
!-----------------------------------------------------------------------------------------
!   Applying the Sheppad filter every 30 numerical iterations
!-----------------------------------------------------------------------------------------

  if (mod(itimestep,30).eq.0) then

    do k=1,ntotal
      i=neighbours(k,1)
      sum_kernel(k)=0.
      niac=2

      do while (neighbours(i,niac).ne.0)
        j=neighbours(i,niac)
        sum_kernel(k) = sum_kernel(k) + w(i,niac)*mass(j)/rho(j)
        niac=niac+1
      end do
    end do

    do k=1,ntotal
      i=neighbours(k,1)
      rho(k)=0.
      niac=2

      do while (neighbours(i,niac).ne.0)
        j=neighbours(i,niac)
        rho(k) = rho(k) + mass(j)* w(i,niac)/sum_kernel(k)
        niac=niac+1
      end do

    end do
```

```
    end if


  ! Storing the results in output files

    aux=itimestep
    if ((itimestep.eq.1).or.(mod(itimestep,recording_step).eq.0)) then

      write(aux2,*) itimestep           ! Converting an integer to a character

      if (itimestep.eq.1) then
        funit=itimestep+131000
      else
        funit=int(itimestep/recording_step)+131000
      end if

      if (aux.lt.10)  filename='../output/data/density/000000'//trim(adjustl(aux2))//'.dat'
      if                                          ((aux.ge.10).and.(aux.lt.100))
filename='../output/data/density/00000'//trim(adjustl(aux2))//'.dat'
      if                                         ((aux.ge.100).and.(aux.lt.1000))
filename='../output/data/density/0000'//trim(adjustl(aux2))//'.dat'
      if                                        ((aux.ge.1000).and.(aux.lt.10000))
filename='../output/data/density/000'//trim(adjustl(aux2))//'.dat'
      if                                       ((aux.ge.10000).and.(aux.lt.100000))
filename='../output/data/density/00'//trim(adjustl(aux2))//'.dat'
      if                                      ((aux.ge.100000).and.(aux.lt.1000000))
filename='../output/data/density/0'//trim(adjustl(aux2))//'.dat'
      if (aux.eq.1000000) filename='../output/data/density'//trim(adjustl(aux2))//'.dat'

      open(unit=funit,file=filename)
      do k= 1, ntotal
        write (funit,221) rho(k), itype(k)
      end do

 221   format (D21.14,1x, I8)

      close (funit)

    end if

end subroutine continuity_density



subroutine pressure(itimestep, ntotal, x, rho, c_sound, mass, viscosity, p, w, itype,
neighbours, surface_particles,K_scale,hsml,Rp_ij)

!----------------------------------------------------------------------------------------------
!  In this subroutine, the pressure applied to ech fluid particle is predicted.
!  ntotal -- number of particles                                                    [in]
!  rho -- density                                                                   [in]
```

```
!  p -- absolute pressure                                              [out]
!  c_sound  -- sound velocity in fluid                                 [out]
!----------------------------------------------------------------------------

    implicit none
    include 'param.inc'

    integer  i,j,k,d,t, itimestep, niac, ntotal, itype(maxn), neighbours(maxn,n_neighbours),
surface_particles(maxn)
    double precision rho(maxn), mass(maxn), p(maxn), x(maxn,dim),
w(maxn,n_neighbours), hsml(maxn), K_scale
    double precision viscosity(maxn), Rp_ij(maxn,n_neighbours), Rpi, Rpj
    double precision c_sound,gamma, rho0, hydrostatic_pressure,
reference_particle_pressure
    double precision grav, b, p_tait, aux_x, aux_z, planar_distance,
shortest_planar_distance, cutoff

    rho0=1000. ! density of fluid at rest
    grav= 9.81  ! gravity magnitude
    gamma=7.    ! exponent in Tait equation
    b=(c_sound**2)*rho0/gamma   ! parameter in Tait equation

    if (itimestep.gt.1) then

     rewind(75)

     do i=1,ntotal
       shortest_planar_distance = 100.
       cutoff = K_scale*hsml(i)
       planar_distance = 100.

       do k =1, ntotal

        if (surface_particles(k).NE.0) then

       !------ Finding the free surface particle closest to the fixed particle under study

         aux_x = x(i,1) - x(surface_particles(k),1)
         aux_z = x(i,3)- x(surface_particles(k),3)
         planar_distance = abs(aux_x**2 + aux_z**2)

         if (planar_distance.le.cutoff) then

          if  (planar_distance.LE.shortest_planar_distance)  shortest_planar_distance  =
planar_distance

            if  (  x(surface_particles(k),2).ge.x(i,2)  )      reference_particle_pressure  =
surface_particles(k)

         end if
```

```
        end if

      end do
        hydrostatic_pressure = rho(i)* grav *(x(reference_particle_pressure,2)-x(i,2))
        p_tait = b*  (((rho(i)/rho0)**gamma)-1.)  ! Dynamic pressure
        p(i) = hydrostatic_pressure + p_tait    ! Hydrostatic pressure + Dynamic pressure
(estimated by Tait eqution)

    end do

  !-----------------------------------------------------------------------------------------------
  ! FOR THE CALCULATION OF ARTIFICIAL PRESSURE
  ! Monaghan. SPH Without a Tensile Instability. Journal of Computational Physics
  ! Lobovsky L., Kren J. Smoothed particle hydrodynamics modelling of fluids and
  ! solids.  Applied and Computational Mechanics.
  !-------------------------------------------------------------------------------------
    if (Artificial_press.eq.'YES') then

      do i=1, ntotal
        niac=2

        if (p(i).lt.0.) then
          Rpi =(lambda*abs(p(i)))/(rho(i)*rho(i))
        else
          Rpi = 0.
        end if

        do while (neighbours(i,niac).ne.0)
          j = neighbours(i,niac)
          if (p(j).lt.0.) then
            Rpj =(lambda*abs(p(j)))/(rho(j)*rho(j))
          else
            Rpj = 0.
          end if
            Rp_ij(i,niac) = Rpi + Rpj
            niac = niac + 1
        end do
      end do
    end if

    do t= 1,ntotal
      write(75,55) t, rho(t), mass(t), p(t)
    end do

55    format(I8,3(1x,D21.14))

   end if

end subroutine pressure
```

```
subroutine internal_force(itimestep,ntotal,hsml,mass,v,neighbours,r,factor,p, c_sound,
pressure_force,viscous_force,rho,viscosity,dx,dy,dz,dwdx,dwdy,dwdz,dwdr,
recording_step, Rp_ij, Fij_art_pressure)

!----------------------------------------------------------------------------------------------------
! Subroutine to calculate the internal forces on the right hand side of the Navier-
! Stokes equations, including the pressure gradient and viscous forces, used by the
! time integration
!
!  itimestep -- current timestep number                                        [in]
!  ntotal -- number of particles                                               [in]
!  hsml -- smoothing length                                                    [in]
!  mass  -- particle masses                                                    [in]
!  niac  -- number of interaction pairs                                        [in]
!  rho   -- density                                                            [in]
!  viscosity -- dynamic viscosity                                              [in]
!  dwdx -- derivative of kernel with respect to x                              [in]
!  dwdy -- derivative of kernel with respect to y                              [in]
!  dwdz -- derivative of kernel with respect to z                              [in]
!  Fij_art_pressure -- force related to the artificial pressure                [in]
!----------------------------------------------------------------------------------------------------

   implicit none
   include 'param.inc'

   integer i, j, k, d, itimestep, ntotal,niac, funit, funit2, filename, recording_step,
neighbours(maxn,n_neighbours)
   double precision hsml(maxn), mass(maxn), v(maxn,dim),rho(maxn),
dwdx(maxn,n_neighbours), dwdy(maxn,n_neighbours), dwdz(maxn,n_neighbours),
dwdr(maxn,n_neighbours), r(maxn,n_neighbours), p(maxn),
viscous_force(maxn,dim),artificial_viscous_force(maxn,dim),
pressure_force(maxn,dim), viscosity(maxn),factor, dx(maxn,n_neighbours),
dy(maxn,n_neighbours), dz(maxn,n_neighbours), den_CSPMx(maxn),
den_CSPMy(maxn), den_CSPMz(maxn), c_sound
   double precision q, aux, aux_3, aux_4, aux_5, r2, c_average, amu, pi_visc
   double precision Fij_art_pressure(maxn,n_neighbours), Rp_ij(maxn,n_neighbours),
aux_art_p_x, aux_art_p_y, aux_art_p_z, Force_art_pres(maxn,dim)
   character(len=25) aux2
   character(len=45) pressure_name, viscosity_name

   do k=1,ntotal
    niac=2
    i = neighbours(k,1)
    viscous_force(i,1)=0.
    viscous_force(i,2)=0.
    viscous_force(i,3)=0.

    do while (neighbours(i,niac).ne.0)
      j = neighbours(i,niac)
```

```
        aux = ( (viscosity(j)+viscosity(i)) * mass(j) )/ ( (rho(i)*rho(j)* (r(i,niac)**2) + (0.01
* hsml(i)**2) ) )
        aux_3 = r(i,niac)*dwdr(i,niac) * (v(j,1)-v(i,1))
        aux_4 =  r(i,niac)*dwdr(i,niac) * (v(j,2)-v(i,2))
        aux_5=  r(i,niac)*dwdr(i,niac) * (v(j,3)-v(i,3))

        viscous_force(i,1)=viscous_force(i,1) + aux * aux_3
        viscous_force(i,2)=viscous_force(i,2) + aux * aux_4
        viscous_force(i,3)=viscous_force(i,3) + aux * aux_5
        niac=niac+1

      end do

    end do

    c_average= c_sound

    do k=1,ntotal
      niac=2
      i = neighbours(k,1)
      artificial_viscous_force(i,1)=0.
      artificial_viscous_force(i,2)=0.
      artificial_viscous_force(i,3)=0.

      do while (neighbours(i,niac).ne.0)
        j = neighbours(i,niac)

        aux_3 = (v(j,1)-v(i,1)) * dx(i,niac) +  (v(j,2)-v(i,2)) * dy(i,niac)  +  (v(j,3)-v(i,3)) *
dz(i,niac)
        r2 = dx(i,niac)**2 + dy(i,niac)**2  + dz(i,niac)**2
        amu = (hsml(i)* aux_3)/ (r2 + 0.01*hsml(i)**2)

        pi_visc = - alpha * c_average * amu / rho(i)

        artificial_viscous_force(i,1)=artificial_viscous_force(i,1) - pi_visc * dwdx(i,niac) *
mass(j)
        artificial_viscous_force(i,2)=artificial_viscous_force(i,2) - pi_visc * dwdy(i,niac) *
mass(j)
        artificial_viscous_force(i,3)=artificial_viscous_force(i,3) - pi_visc * dwdz(i,niac) *
mass(j)
       niac=niac+1

      end do

    end do

!------------------------------------------------------------------------------------------------
!-----------------PRESSURE FORCES AND ARTIFICIAL PRESSURE--------------------
!------------------------------------------------------------------------------------------------
   do k=1, ntotal
     niac=2
```

```
i = neighbours(k,1)
pressure_force(i,1)=0.
pressure_force(i,2)=0.
pressure_force(i,3)=0.
den_CSPMx(i)=0.
den_CSPMy(i)=0.
den_CSPMz(i)=0.
Force_art_pres(i,1) = 0.
Force_art_pres(i,2) = 0.
Force_art_pres(i,3) = 0.

do while (neighbours(i,niac).ne.0)
  j = neighbours(i,niac)
  aux_3 = - ( (p(i)/rho(i)**2.) + (p(j)/rho(j)**2) ) * dwdx(i,niac) * mass(j)
  aux_4 = - ( (p(i)/rho(i)**2.) + (p(j)/rho(j)**2) ) * dwdy(i,niac) * mass(j)
  aux_5 = - ( (p(i)/rho(i)**2.) + (p(j)/rho(j)**2) ) * dwdz(i,niac) * mass(j)

  den_CSPMx(i)= den_CSPMx(i)+ dx(i,niac)* (mass(j)/rho(j)) * dwdx(i,niac) !
FOR CSPM CORRECTION
  den_CSPMy(i)= den_CSPMy(i)+ dy(i,niac)* (mass(j)/rho(j)) * dwdy(i,niac)
  den_CSPMz(i)= den_CSPMz(i)+ dz(i,niac)* (mass(j)/rho(j)) * dwdz(i,niac)

  pressure_force(i,1)=pressure_force(i,1) + aux_3
  pressure_force(i,2)=pressure_force(i,2) + aux_4
  pressure_force(i,3)=pressure_force(i,3) + aux_5

 !-----------------------------------------------------------------------------------------------------
 !--------------------------------- A R T I F I C I A L  P R E S S U R E------------------------
 !-----------------------------------------------------------------------------------------------------
  if (Artificial_press.eq.'YES') then
   aux_art_p_x = - (Rp_ij(i,niac))*((Fij_art_pressure(i,niac))**4.)*dwdx(i,niac) *
mass(j)
   aux_art_p_y = - (Rp_ij(i,niac))*((Fij_art_pressure(i,niac))**4.)*dwdy(i,niac) *
mass(j)
   aux_art_p_z = - (Rp_ij(i,niac))*((Fij_art_pressure(i,niac))**4.)*dwdz(i,niac) *
mass(j)

   Force_art_pres(i,1) = Force_art_pres(i,1) + aux_art_p_x
   Force_art_pres(i,2) = Force_art_pres(i,2) + aux_art_p_y
   Force_art_pres(i,3) = Force_art_pres(i,3) + aux_art_p_z
  end if
 !-----------------------------------------------------------------------------------------------------

  niac=niac+1

end do

pressure_force(i,1)=pressure_force(i,1)/den_CSPMx(i) + artificial_viscous_force(i,1)
+ Force_art_pres(i,1)
```

```
   pressure_force(i,2)=pressure_force(i,2)/den_CSPMy(i) + artificial_viscous_force(i,2)
+ Force_art_pres(i,2)
   pressure_force(i,3)=pressure_force(i,3)/den_CSPMz(i) + artificial_viscous_force(i,3)
+ Force_art_pres(i,3)

 end do

 open(71,file='../output/pressure_force.dat')
 rewind(71)
 write(71,*) ' ITERATION NUMBER = ', itimestep
 do k=1,ntotal
  write(71,*)  pressure_force(i,1), pressure_force(i,2), pressure_force(i,3)
 end do

 close (71)

! Storing the results in output files

 filename=itimestep
 if ((itimestep.eq.1).or.(mod(itimestep,recording_step).eq.0)) then

  write(aux2,*) itimestep          ! Converting an integer to a character

  if (itimestep.eq.1) then
   funit=itimestep+3100
   funit2=itimestep+60100
  else
    funit=int(itimestep/recording_step)+3100
    funit2=int(itimestep/recording_step)+61000
   end if

   if                                                        (filename.lt.10)
pressure_name='../output/data/pressure_force/000000'//trim(adjustl(aux2))//'.dat'
   if                                    ((filename.GE.10).and.(filename.lt.100))
pressure_name='../output/data/pressure_force/00000'//trim(adjustl(aux2))//'.dat'
   if                                   ((filename.ge.100).and.(filename.lt.1000))
pressure_name='../output/data/pressure_force/0000'//trim(adjustl(aux2))//'.dat'
   if                                 ((filename.ge.1000).and.(filename.lt.10000))
pressure_name='../output/data/pressure_force/000'//trim(adjustl(aux2))//'.dat'
   if                               ((filename.ge.10000).and.(filename.lt.100000))
pressure_name='../output/data/pressure_force/00'//trim(adjustl(aux2))//'.dat'
   if                             ((filename.ge.100000).and.(filename.lt.1000000))
pressure_name='../output/data/pressure_force/0'//trim(adjustl(aux2))//'.dat'
   if                                                   (filename.eq.1000000)
pressure_name='../output/data/pressure_force'//trim(adjustl(aux2))//'.dat'

   open(unit=funit,file=pressure_name)
   do i = 1, ntotal
    write (funit,220) i, pressure_force(i,1), pressure_force(i,2), pressure_force(i,3)
   end do
```

```
!-----------------------------------------------------------------------------------------------------

    if                                                              (filename.lt.10)
viscosity_name='../output/data/viscous_force/000000'//trim(adjustl(aux2))//'.dat'
    if                                          ((filename.ge.10).and.(filename.lt.100))
viscosity_name='../output/data/viscous_force/00000'//trim(adjustl(aux2))//'.dat'
    if                                        ((filename.ge.100).and.(filename.lt.1000))
viscosity_name='../output/data/viscous_force/0000'//trim(adjustl(aux2))//'.dat'
    if                                      ((filename.ge.1000).and.(filename.lt.10000))
viscosity_name='../output/data/viscous_force/000'//trim(adjustl(aux2))//'.dat'
    if                                    ((filename.ge.10000).and.(filename.lt.100000))
viscosity_name='../output/data/viscous_force/00'//trim(adjustl(aux2))//'.dat'
    if                                  ((filename.ge.100000).and.(filename.lt.1000000))
viscosity_name='../output/data/viscous_force/0'//trim(adjustl(aux2))//'.dat'
    if                                                         (filename.eq.1000000)
viscosity_name='../output/data/viscous_force'//trim(adjustl(aux2))//'.dat'

    open(unit=funit2,file=viscosity_name)
    do i = 1, ntotal
      write (funit2,220) i, viscous_force(i,1), viscous_force(i,2), viscous_force(i,3)
    end do

 220 format (I8, 3(1x,d21.14))
    close (funit)
    close(funit2)

  end if

end subroutine internal_force


subroutine external_force(ntotal,gravity)

!-----------------------------------------------------------------------------------------------------
!   Subroutine to calculate the external forces, in this case, the gravitational forces.
!-----------------------------------------------------------------------------------------------------

   implicit none
   include 'param.inc'

   integer i, ntotal
   double precision  gravity(maxn,dim)

   do i=1,ntotal
     gravity(i,1)=0.
     gravity(i,2) =-9.81
     gravity(i,3) = 0.
   end do
```

```
end subroutine external_force



subroutine acceleration(itimestep,mass, ntotal, pressure_force, viscous_force, gravity,
accel, recording_step)

   implicit none
   include 'param.inc'

   integer ntotal,i, itimestep, aux, funit, recording_step
   double precision pressure_force(maxn,dim), viscous_force(maxn,dim),
gravity(maxn,dim), accel(maxn,dim), mass(maxn)
   character(len=25) aux2
   character(len=45) filename

   do i=1,ntotal
    accel(i,1) = -pressure_force(i,1)+gravity(i,1)+viscous_force(i,1)
    accel(i,2) = -pressure_force(i,2)+gravity(i,2)+viscous_force(i,2)
    accel(i,3) = -pressure_force(i,3)+gravity(i,3)+viscous_force(i,3)
   end do

   aux=itimestep
   if ((itimestep.eq.1).or.(mod(itimestep,recording_step).eq.0)) then
    write(aux2,*) itimestep           ! Converting an integer to a character

    if (itimestep.eq.1) then
     funit=itimestep+100
    else
     funit=int(itimestep/recording_step)+100000
    end if

    if (aux.lt.10)
filename='../output/data/acceleration/000000'//trim(adjustl(aux2))//'.dat'
    if ((aux.ge.10).and.(aux.lt.100))
filename='../output/data/acceleration/00000'//trim(adjustl(aux2))//'.dat'
    if ((aux.ge.100).and.(aux.lt.1000))
filename='../output/data/acceleration/0000'//trim(adjustl(aux2))//'.dat'
    if ((aux.ge.1000).and.(aux.lt.10000))
filename='../output/data/acceleration/000'//trim(adjustl(aux2))//'.dat'
    if ((aux.ge.10000).and.(aux.lt.100000))
filename='../output/data/acceleration/00'//trim(adjustl(aux2))//'.dat'
    if ((aux.ge.100000).and.(aux.lt.1000000))
filename='../output/data/acceleration/0'//trim(adjustl(aux2))//'.dat'
    if (aux.eq.1000000) filename='../output/data/acceleration'//trim(adjustl(aux2))//'.dat'

    open(unit=funit,file=filename)
    do i = 1, ntotal
     write (funit,221) i, accel(i,1), accel(i,2), accel(i,3)
    end do
221  format (I8, 3(1x,D21.14))
```

```
      close (funit)

    end if

end subroutine acceleration


subroutine XSPH(ntotal,mass,neighbours,w,rho,v,v_xsph)

!------------------------------------------------------------------------------
! Subroutine to calculate the average velocity to correct velocity for preventing
! particle penetration
!  ntotal -- Number of particles                                          [in]
!  mass -- Particle masses                                                [in]
!  w -- Kernel for all interaction pairs                                  [in]
!  v -- Velocity of each particle                                         [in]
!  v_xsph -- Corrected particle velocity (XSPH)                          [out]
!  rho -- Density of each particle                                        [in]
!------------------------------------------------------------------------------

    implicit none
    include 'param.inc'

    integer i, j, k, d, ntotal,niac, neighbours(maxn,n_neighbours)
    double precision    mass(maxn), w(maxn,n_neighbours), rho(maxn), v(maxn,dim),
v_xsph(maxn,dim), rho_medio, dv(dim), epsilon

!   epsilon --- a small constant chosen by experience, may lead to instability.

    epsilon = 0.40

    do i = 1, ntotal
      do d = 1, 3
        v_xsph(i,d) = 0.
      end do
    end do

    do k=1,ntotal
      niac = 2
      i = neighbours(k,1)

      do while (neighbours(i,niac).ne.0)
        j=neighbours(i,niac)

        do d=1,3
          rho_medio = (rho(i)+rho(j))/2.
          dv(d) = v(j,d) - v(i,d)
          v_xsph (i,d) = v_xsph(i,d) + mass(j)*dv(d)*w(i,niac)/rho_medio
        end do
        niac=niac+1
```

```
      end do

    end do

    do k = 1, ntotal
      do d = 1,dim
        v_xsph(i,d) = epsilon * v_xsph (i,d)
      end do
    end do

end subroutine XSPH



subroutine time_integration(itimestep,accel,ntotal,dt,x,v,v_xsph,recording_step)

!-----------------------------------------------------------------------------------------------
!  ntotal -- Total particle number                                              [in]
!  itimestep -- Current iteration                                               [in]
!  x --  Coordinates of particles                                           [in/out]
!  v --    Velocities of particles                                          [in/out]
!  mass --  Mass of particles                                                   [in]
!  rho --    Densities of particles                                         [in/out]
!  p --    Pressure  of particles                                           [in/out]
!  hsml --   Smoothing lengths of particles                                 [in/out]
!  dt -- Timestep                                                               [in]
!-----------------------------------------------------------------------------------------------

    implicit none
    include 'param.inc'

    integer ntotal,itype(maxn),itimestep,i,d,recording_step
    double  precision  x(maxn,dim),  v(maxn,dim),  xpos(maxn,4),  auxv(maxn,4),
accel(maxn,dim), dt, v_xsph (maxn,dim)

    rewind(45)
    rewind(46)

    do i=1, ntotal
      read(45,*)  (auxv(i,d), d=1,4)
      read(46,*)  (xpos(i,d), d=1,4)
    end do

    open(40,file="../output/initial_velocity.dat")
    open(41,file="../output/initial_position.dat")
    rewind(40)
    rewind(41)

    do i=1, ntotal
      write(40,54) int(auxv(i,1)), (auxv(i,d), d=2,4)
```

```
   write(41,54) int(xpos(i,1)), (xpos(i,d), d=2,4)
  end do

  close(40)
  close(41)

  if (XSPH_correction.eq."YES") then
   do i=1, ntotal
    v(i,1) = auxv(i,2) + accel(i,1)*dt + v_xsph(i,1)    ! Euler's Integration Method
(efficient for small enough time steps)
    v(i,2) = auxv(i,3) + accel(i,2)*dt + v_xsph(i,2)
    v(i,3) = auxv(i,4) + accel(i,3)*dt + v_xsph(i,3)
    x(i,1) = xpos(i,2) + v(i,1) *dt
    x(i,2) = xpos(i,3) + v(i,2) *dt
    x(i,3) = xpos(i,4) + v(i,3) *dt
   end do
  else
   do i=1, ntotal
    v(i,1) = auxv(i,2) + accel(i,1)*dt
    v(i,2) = auxv(i,3) + accel(i,2)*dt
    v(i,3) = auxv(i,4) + accel(i,3)*dt
    x(i,1) = xpos(i,2) + v(i,1) *dt
    x(i,2) = xpos(i,3) + v(i,2) *dt
    x(i,3) = xpos(i,4) + v(i,3) *dt
   end do
  end if

  rewind(46)
  rewind(45)

  do i=1, ntotal
   write(45,54) i, (v(i,d), d = 1,dim)
   write(46,54) i, (x(i,d), d = 1,dim)
  end do

54  format(I8, 3(1x,D21.14))

 end subroutine time_integration


subroutine free_surface(itimestep,ntotal,recording_step,surface_particles,p,x)

!-----------------------------------------------------------------------------------------------
! This subroutine creates output files with the updated position of the free surface particles
!-----------------------------------------------------------------------------------------------

  implicit none
  include 'param.inc'

  integer      i, ntotal, itimestep, recording_step, funit, aux, surface_particles(maxn)
```

```
    double precision p(maxn), x(maxn,dim)
    character(len=25) aux2
    character(len=45) filename

    aux = itimestep
    if ((itimestep.eq.1).or.(mod(itimestep,recording_step).eq.0)) then
     write(aux2,*) itimestep              ! Converting an integer to a character

     if (itimestep.eq.1) then
       funit=itimestep+15000
     else
       funit=int(itimestep/recording_step)+15001
     end if

     if                                                              (aux.lt.10)
filename='../output/data/free_surface/000000'//trim(adjustl(aux2))//'.dat'
     if                                            ((aux.ge.10).and.(aux.lt.100))
filename='../output/data/free_surface/00000'//trim(adjustl(aux2))//'.dat'
     if                                          ((aux.ge.100).and.(aux.lt.1000))
filename='../output/data/free_surface/0000'//trim(adjustl(aux2))//'.dat'
     if                                        ((aux.ge.1000).and.(aux.lt.10000))
filename='../output/data/free_surface/000'//trim(adjustl(aux2))//'.dat'
     if                                      ((aux.ge.10000).and.(aux.lt.100000))
filename='../output/data/free_surface/00'//trim(adjustl(aux2))//'.dat'
     if                                    ((aux.ge.100000).and.(aux.lt.1000000))
filename='../output/data/free_surface/0'//trim(adjustl(aux2))//'.dat'
     if (aux.eq.1000000) filename='../output/data/free_surface'//trim(adjustl(aux2))//'.dat'

     open(unit=funit,file=filename)
     do i = 1, ntotal
       if (surface_particles(i).ne.0) then
         write (funit,225)  i, x(i,1), x(i,2), x(i,3)   ! file that monitors free surface particles
       end if
     end do
 225  format(I8, 3(1x,D21.14))

     close (funit)

    end if

end subroutine free_surface



subroutine collision_test(ntotal,radius,itimestep,dt,CR,friction_coef)

    implicit none
    include 'param.inc'

    integer ntotal, itimestep, n_part, i, j, aux, ncol, plane_col, plane_col2, plane_col3,
plane, cont, d
```

```
   double precision radius,dt, x(maxn,dim),v(maxn,dim)
   double precision aux_x, aux_y, aux_z, dx , dy, dz, absolute_value_a, num, den,
aux1_s, aux2_s, aux3_s
   double precision esc_1, esc_2, esc_3, esc_4, esc_5, esc_6, num_escalar, den_escalar,
minor_d, minor_d2, minor_d3
   double precision P_0(6,4), P_I(6,4), PIP0(6,4), a(3), s(6), points_distance(6),
X_0(6,dim), n(6,dim), reflected_position(3), new_C0(3), CR, friction_coef
   integer,dimension(:,:),allocatable::planes
   double precision,dimension(:,:), allocatable:: position_t0, position_t1, velocity_t1,
distance
   character(len=3) stop_test
   character(len=30) planecol, planecol2, planecol3

   allocate(position_t0(ntotal,4),        position_t1(ntotal,4),        velocity_t1(ntotal,4),
distance(ntotal,7), planes(ntotal,7))

   open(40,file="../output/initial_velocity.dat")
   open(41,file="../output/initial_position.dat")

   rewind(40)  ! velocities at the beginning of the iteration
   rewind(41)  ! positions at the beginning of the iteration
   rewind(45)  ! velocities at the end of the iteration
   rewind(46)  ! positions at the end of the iteration
   rewind(445) ! normal vectors to the walls
   rewind(450) ! points belonging to the walls

   do n_part=1,ntotal
    read (41,*) (position_t0(n_part,j),j=1,4)
  end do
  close(41)

  do n_part=1,ntotal
   read (46,*) (position_t1(n_part,j),j=1,4)
  end do

  do n_part=1,ntotal
   read (45,*) (velocity_t1(n_part,j),j=1,4)
  end do

  do i=1,6
   read (445,*) (n(i,j),j=1,dim)
  end do

  do i=1,6
   read(450,*) (X_0(i,j),j=1,dim)
  end do

!----------------------------------------------------------------------------------------------------
! Calculating the distance between the center of the particle in the initial position "C0"
! and in the final position "C1" and the planes
```

```
!-----------------------------------------------------------------------------------------------------------
  stop_test="NO"  !---- Assumption that there were collisions in the first numerical
iteration

  do n_part=1,ntotal
    cont=0
    distance(n_part,1)=n_part

!-----------------------------------------------------------------------------------------------------------
! Checking the possibilities of collisions for the particle "n_part"
!-----------------------------------------------------------------------------------------------------------
    do while (stop_test.eq."NO")  !!!AAA

! Calculations of the distance from the center "C1" of the particle
! "n_part" at the end point of its trajectory, after each possible collision, and each active
! plane:

!--- Distance to plane 01 (column 02 of the distance matrix):  [RIGHT PLANE]

    if(  (n(1,1).ne.0).or.(n(1,2).ne.0).or.(n(1,3).ne.0)  ) then
      aux_x=n(1,1)*(position_t1(n_part,2)-X_0(1,1))
      aux_y=n(1,2)*(position_t1(n_part,3)-X_0(1,2))
      aux_z=n(1,3)*(position_t1(n_part,4)-X_0(1,3))

      distance(n_part,2)=aux_x+aux_y+aux_z

    else
       distance(n_part,2)=99999.
    end if

!--- Distance to plane 02 (column 03 of the distance matrix):  [LEFT PLANE]
    if ( (n(2,1).ne.0).or.(n(2,2).ne.0).or.(n(2,3).ne.0 ) ) then
       aux_x=n(2,1)*(position_t1(n_part,2)-X_0(2,1))
       aux_y=n(2,2)*(position_t1(n_part,3)-X_0(2,2))
       aux_z=n(2,3)*(position_t1(n_part,4)-X_0(2,3))

      distance(n_part,3)=aux_x+aux_y+aux_z

    else
      distance(n_part,3)=99999.
    end if

    !--- Distance to plane 03 (column 04 of the distance matrix):  [LOWER PLANE]

    if ( (n(3,1).ne.0).or.(n(3,2).ne.0).or.(n(3,3).ne.0 ) ) then
      aux_x=n(3,1)*(position_t1(n_part,2)-X_0(3,1))
      aux_y=n(3,2)*(position_t1(n_part,3)-X_0(3,2))
      aux_z=n(3,3)*(position_t1(n_part,4)-X_0(3,3))

      distance(n_part,4)=aux_x+aux_y+aux_z
```

```
    else
     distance(n_part,4)=99999.
    end if

!--- Distance to plane 04 (column 05 of the distance matrix): [UPPER PLANE]

    if ( (n(4,1).ne.0).or.(n(4,2).ne.0).or.(n(4,3).ne.0 ) ) then
     aux_x=n(4,1)*(position_t1(n_part,2)-X_0(4,1))
     aux_y=n(4,2)*(position_t1(n_part,3)-X_0(4,2))
     aux_z=n(4,3)*(position_t1(n_part,4)-X_0(4,3))

     distance(n_part,5)=aux_x+aux_y+aux_z

    else
     distance(n_part,5)=99999.
    end if

!--- Distance to plane 05 (column 06 of the distance matrix): [BACK PLANE]

    if ( (n(5,1).ne.0).or.(n(5,2).ne.0).or.(n(5,3).ne.0 ) ) then
     aux_x=n(5,1)*(position_t1(n_part,2)-X_0(5,1))
     aux_y=n(5,2)*(position_t1(n_part,3)-X_0(5,2))
     aux_z=n(5,3)*(position_t1(n_part,4)-X_0(5,3))

    distance(n_part,6)=aux_x+aux_y+aux_z

    else
     distance(n_part,6)=99999.
    end if

!--- Distance to plane 06 (column 05 of the distance matrix): [FRONTAL PLANE]

    if ( (n(6,1).ne.0).or.(n(6,2).ne.0).or.(n(6,3).ne.0 ) ) then
     aux_x=n(6,1)*(position_t1(n_part,2)-X_0(6,1))
     aux_y=n(6,2)*(position_t1(n_part,3)-X_0(6,2))
     aux_z=n(6,3)*(position_t1(n_part,4)-X_0(6,3))

     distance(n_part,7)=aux_x+aux_y+aux_z

    else
     distance(n_part,7)=99999.
    end if

! ---- Initializing the "planes" matrix with zeros:

    planes(n_part,1)=n_part        !The 1st column of the "planes" matrix identifies the
particle "n_part"
    do j=2,7
     planes(n_part,j)=0      !Assuming initially that there is no collision against any plane
```

```
    end do

!------------------------------------------------------------------------------------------
! Performing the collision test, detecting and storing the planes against which collisions
! may have occurred
! Criterion: distances from the center of the particle at t0+dt and the plane considered
!------------------------------------------------------------------------------------------
   do j=2,7

   ! If there is a possibility of collision against any plane, the position of the "planes" array
relative to the respective collision plane receives the value "one"

    if (distance(n_part,j).GE.0.AND.distance(n_part,j).LE.radius) then
     planes(n_part,j)=1
    else
     if (distance(n_part,j).LT.0) then
      planes(n_part,j)=1
     end if
    end if
   end do

!------------------------------------------------------------------------------------------
! If there was a collision of the particle, it is necessary to check if it will still continue to
! collide with the planes within the planes within the time step
!------------------------------------------------------------------------------------------
   stop_test="YES"

   j=2
   do while (stop_test.eq."YES".and.j.LE.7)
    if (planes(n_part,j).eq.1) then       ! If true, need to check for new collisions
     if (stop_test.eq."YES") then
      cont=cont+1                    ! Record the possible number of collisions
     end if
      stop_test="NO"
    else
     j=j+1
    end if
   end do

!------------------------------------------------------------------------------------------
! Calculation of the "P_0" points of the "n_part" particle in relation to each possible
! collision plane
! Calculations performed only if there is a possibility of collision of the particles against
! one or more planes
!------------------------------------------------------------------------------------------

   if (stop_test.eq."NO") then       !!!!B

    open(unit=773,file='../output/P_0.dat')   ! File containing the coordinates of point P_0
(closest to the collision plane)
```

```
rewind(773)                    !da partícula
write (773,*) n_part
do i=1,6                 ! Initializing the matrix P_0 (referring to the particle "n_part")
 P_0(i,1)=i
  do j=2,4
   P_0(i,j)=0.
  end do      !Columns 2 to 4 of the matrix are set to "zero"
end do        !It is assumed that there was no collision against any plane

!-----------------------------------------------------------------------------------------
! Determining P_0 (position of the particle point that will touch each of the possible
! collision planes). Only if there is a possibility of collision.
!-----------------------------------------------------------------------------------------

do i=2,7
  aux=planes(n_part,i)

  if (aux.ne.0) then

    if (i.eq.2) then      ! P_0 relative to plane 01 [Right plane]
     P_0(i-1,2)=position_t0(n_part,2)+radius
     P_0(i-1,3)=position_t0(n_part,3)
     P_0(i-1,4)=position_t0(n_part,4)
    else
     if (i.eq.3) then     ! P_0 relative to plane 02 [Left plane]
       P_0(i-1,2)=position_t0(n_part,2)-radius
       P_0(i-1,3)=position_t0(n_part,3)
       P_0(i-1,4)=position_t0(n_part,4)
     else
      if (i.eq.4) then  ! P_0 relative to plane 03 [Bottom]
        P_0(i-1,2)=position_t0(n_part,2)
        P_0(i-1,3)=position_t0(n_part,3)-radius
        P_0(i-1,4)=position_t0(n_part,4)
      else
       if (i.eq.5) then   ! P_0 relative to plane 04 [Upper]
         P_0(i-1,2)=position_t0(n_part,2)
         P_0(i-1,3)=position_t0(n_part,3)+radius
         P_0(i-1,4)=position_t0(n_part,4)
       else
        if (i.eq.6) then  !P_0 relative to plane 05 [Back plane]
          P_0(i-1,2)=position_t0(n_part,2)
          P_0(i-1,3)=position_t0(n_part,3)
          P_0(i-1,4)=position_t0(n_part,4)-radius
        else
         if (i.eq.7) then    !P_0 relative to plane 06  ! [Frontal plane]
           P_0(i-1,2)=position_t0(n_part,2)
           P_0(i-1,3)=position_t0(n_part,3)
           P_0(i-1,4)=position_t0(n_part,4)+radius
         end if
        end if
```

```
            end if
          end if
        end if
      end if

    end if

  end do

  do i=1,6
    write(773,27) (P_0(i,j),j=1,4)      ! Writing P_0 to an output file
27  format (4(2x,E13.6))
  end do
  close(773)

!---------------------------------------------------------------------------------------------
! Finding the directing vector "a" of the line cointaining the trajectory of "P_0"
! Calculate "a" from the positions of the centres of the particles at the beginning (C0) and
! at the end of the trajectory (C1)
! It is known that the lines containing points P_0 and
! P_1 (line 1) and !the one containing points C0 and C1 are parallel
!---------------------------------------------------------------------------------------------
  dx=position_t1(n_part,2)-position_t0(n_part,2)
  dy=position_t1(n_part,3)-position_t0(n_part,3)
  dz=position_t1(n_part,4)-position_t0(n_part,4)
  absolute_value_a=sqrt(dx*dx+dy*dy+dz*dz)

! Obtaining the directing vector of the line that contains the trajectory of the particle until
the collision

  do i=2,4
    a(i-1)=(position_t1(n_part,i)-position_t0(n_part,i))/absolute_value_a
  end do

!---------------------------------------------------------------------------------------------
! Calculating the parameters of the line that intercepts the collision plane - equation of the
! line: P_I = P0 + s*a
!---------------------------------------------------------------------------------------------
  do i=1,6
    s(i)=99999.    ! If there is no collision against the plane i, the value of 's' relative to
the collision against that plane will be 99999.
    PIP0(i,1) = i
    PIP0(i,2) = 99999.
    PIP0(i,3) = 99999.
    PIP0(i,4) = 99999.
  end do

!---------------------------------------------------------------------------------------------
!---------- DETERMINATION OF POSSIBLE COLLISION POINTS P_I --------------
```

```
!-------------(OBTAINED FROM THE EQUATION OF THE PARTICLE
TRAJECTORY - EQUATION OF THE STRAIGHT LINE)----------------------------------
!---------------------------------------------------------------------------------------------------

  do i=1,6

   if ((i.eq.1).and.(planes(n_part,i+1).ne.0)) then

     ! ----  Possible collision against the Plane 01 ----

     ! X_0 belongs to the plane and P_0 belongs to the line
     ! Calculating X_0-P_0:
      aux1_s=X_0(1,1)-P_0(i,2)
      aux2_s=X_0(1,2)-P_0(i,3)
      aux3_s=X_0(1,3)-P_0(i,4)
     ! Taking the scalar product between (X_0-P_0) and the normal vector to the plane
      esc_1=(aux1_s)*n(1,1)
      esc_2=(aux2_s)*n(1,2)
      esc_3=(aux3_s)*n(1,3)
      num_escalar=esc_1+esc_2+esc_3
     ! Making the scalar product between the direction vector of the line and the normal
vector to the plane
      esc_4=a(1)*n(1,1)
      esc_5=a(2)*n(1,2)
      esc_6=a(3)*n(1,3)
      den_escalar=esc_4+esc_5+esc_6

      s(1)=num_escalar/den_escalar   !-------- s is the distance between PI and P0

      !----------- Collision point against the plane 1 --------------------
      P_I(i,1)=i
      P_I(i,2) = P_0(i,2) + s(1)*a(1)
      P_I(i,3) = P_0(i,3) + s(1)*a(2)
      P_I(i,4) = P_0(i,4) + s(1)*a(3)

      !------- PI - P0 = PIP0 (TO FIND THE COMPONENTS OF THE COLLISION
DISTANCE AGAINST THE PLANE 1)
      PIP0(i,1) = i
      PIP0(i,2) = (s(1)*a(1))
      PIP0(i,3) = (s(1)*a(2))
      PIP0(i,4) = (s(1)*a(3))

   else
    if ((i.eq.2).and.(planes(n_part,i+1).ne.0)) then

     ! ---- Possible collision against the Plane 02  ----

      ! X_0 belongs to the plane and P_0 belongs to the line
      ! Calculating X_0-P_0:
       aux1_s=X_0(2,1)-P_0(i,2)
```

```
    aux2_s=X_0(2,2)-P_0(i,3)
    aux3_s=X_0(2,3)-P_0(i,4)
  ! Taking the scalar product between (X_0-P_0) and the normal vector to the plane
    esc_1=(aux1_s)*n(2,1)
    esc_2=(aux2_s)*n(2,2)
    esc_3=(aux3_s)*n(2,3)
    num_escalar=esc_1+esc_2+esc_3
  ! Making the scalar product between the direction vector of the line and the normal
vector to the plane
    esc_4=a(1)*n(2,1)
    esc_5=a(2)*n(2,2)
    esc_6=a(3)*n(2,3)
    den_escalar=esc_4+esc_5+esc_6

    s(2)=num_escalar/den_escalar   !-------- s is the distance between PI and P0

  !----------- Collision point against the plane 2 -----------
    P_I(i,1)=i
    P_I(i,2) = P_0(i,2) + s(2)*a(1)
    P_I(i,3) = P_0(i,3) + s(2)*a(2)
    P_I(i,4) = P_0(i,4) + s(2)*a(3)

  ! ---------- PI - P0 = PIP0 (TO FIND THE COMPONENTS OF THE COLLISION
DISTANCE AGAINST THE PLANE 2)
    PIP0(i,1) = i
    PIP0(i,2) = (s(2)*a(1))
    PIP0(i,3) = (s(2)*a(2))
    PIP0(i,4) = (s(2)*a(3))

  else
   if ((i.eq.3).and.(planes(n_part,i+1).ne.0)) then

    ! ---- Possible collision against the Plane 03 ----

    ! X_0 belongs to the plane and P_0 belongs to the line
    ! Calculating X_0-P_0:
      aux1_s=X_0(3,1)-P_0(i,2)
      aux2_s=X_0(3,2)-P_0(i,3)
      aux3_s=X_0(3,3)-P_0(i,4)
    ! Taking the scalar product between (X_0-P_0) and the normal vector to the plane
      esc_1=(aux1_s)*n(3,1)
      esc_2=(aux2_s)*n(3,2)
      esc_3=(aux3_s)*n(3,3)
      num_escalar=esc_1+esc_2+esc_3
    ! Making the scalar product between the direction vector of the line and the normal
vector to the plane
      esc_4=a(1)*n(3,1)
      esc_5=a(2)*n(3,2)
      esc_6=a(3)*n(3,3)
      den_escalar=esc_4+esc_5+esc_6
```

```
    s(3)=num_escalar/den_escalar      !-------- s is the distance between PI and P0

   ! ----------- Collision point against the plane 3 ---------------------
    P_I(i,1)=i
    P_I(i,2) = P_0(i,2) + s(3)*a(1)
    P_I(i,3) = P_0(i,3) + s(3)*a(2)
    P_I(i,4) = P_0(i,4) + s(3)*a(3)

   !--------- PI - P0 = PIP0 (TO FIND THE COMPONENTS OF THE COLLISION
DISTANCE AGAINST THE PLANE 3)
    PIP0(i,1) = i
    PIP0(i,2) = (s(3)*a(1))
    PIP0(i,3) = (s(3)*a(2))
    PIP0(i,3) = (s(3)*a(3))

 else
  if ((i.eq.4).and.(planes(n_part,i+1).ne.0)) then

   ! ---- Possible collision against the Plane 04 ----

   ! X_0 belongs to the plane and P_0 belongs to the line
   ! Calculating X_0-P_0:
     aux1_s=X_0(4,1)-P_0(i,2)
     aux2_s=X_0(4,2)-P_0(i,3)
     aux3_s=X_0(4,3)-P_0(i,4)
   ! Taking the scalar product between (X_0-P_0) and the normal vector to the
plane
     esc_1=(aux1_s)*n(4,1)
     esc_2=(aux2_s)*n(4,2)
     esc_3=(aux3_s)*n(4,3)
     num_escalar=esc_1+esc_2+esc_3
   ! Making the scalar product between the direction vector of the line and the
normal vector to the plane
     esc_4=a(1)*n(4,1)
     esc_5=a(2)*n(4,2)
     esc_6=a(3)*n(4,3)
     den_escalar=esc_4+esc_5+esc_6

    s(4)=num_escalar/den_escalar    !-------- s is the distance between PI and P0

   ! ----------- Collision point against the plane 4 ---------------------
    P_I(i,1)=i
    P_I(i,2) = P_0(i,2) + s(4)*a(1)
    P_I(i,3) = P_0(i,3) + s(4)*a(2)
    P_I(i,4) = P_0(i,4) + s(4)*a(3)

   !---------- PI - P0 = PIP0 (TO FIND THE COMPONENTS OF THE COLLISION
DISTANCE AGAINST THE PLANE 4)
    PIP0(i,1) = i
```

```
PIPO(i,2) = (s(4)*a(1))
PIPO(i,3) = (s(4)*a(2))
PIPO(i,4) = (s(4)*a(3))

else
if ((i.eq.5).and.(planes(n_part,i+1).ne.0)) then

  ! ---- Possible collision against the Plane 05 ----

  ! X_0 belongs to the plane and P_0 belongs to the line
  ! Calculating X_0-P_0:
   aux1_s=X_0(5,1)-P_0(i,2)
   aux2_s=X_0(5,2)-P_0(i,3)
   aux3_s=X_0(5,3)-P_0(i,4)
  ! Taking the scalar product between (X_0-P_0) and the normal vector to the
plane
   esc_1=(aux1_s)*n(5,1)
   esc_2=(aux2_s)*n(5,2)
   esc_3=(aux3_s)*n(5,3)
   num_escalar=esc_1+esc_2+esc_3
  ! Making the scalar product between the direction vector of the line and the
normal vector to the plane
   esc_4=a(1)*n(5,1)
   esc_5=a(2)*n(5,2)
   esc_6=a(3)*n(5,3)
   den_escalar=esc_4+esc_5+esc_6

   s(5)=num_escalar/den_escalar   !-------- s is the distance between PI and P0

  ! ----------- Collision point against the plane 5 --------------------
   P_I(i,1)=i
   P_I(i,2) = P_0(i,2) + s(5)*a(1)
   P_I(i,3) = P_0(i,3) + s(5)*a(2)
   P_I(i,4) = P_0(i,4) + s(5)*a(3)

  !--------- PI - P0 = PIP0 (TO FIND THE COMPONENTS OF THE
COLLISION DISTANCE AGAINST THE PLANE 5)
   PIP0(i,1) = i
   PIP0(i,2) = (s(5)*a(1))
   PIP0(i,3) = (s(5)*a(2))
   PIP0(i,4) = (s(5)*a(3))

 else
  if ((i.EQ.6).AND.(planes(n_part,i+1).NE.0)) then

  ! ---- Possible collision against the Plane 06 ----

  ! X_0 belongs to the plane and P_0 belongs to the line
  ! Calculating X_0-P_0:
    aux1_s=X_0(6,1)-P_0(i,2)
```

```
            aux2_s=X_0(6,2)-P_0(i,3)
            aux3_s=X_0(6,3)-P_0(i,4)
          ! Taking the scalar product between (X_0-P_0) and the normal vector to the
plane
            esc_1=(aux1_s)*n(6,1)
            esc_2=(aux2_s)*n(6,2)
            esc_3=(aux3_s)*n(6,3)
            num_escalar=esc_1+esc_2+esc_3
          ! Making the scalar product between the direction vector of the line and the
normal vector to the plane
            esc_4=a(1)*n(6,1)
            esc_5=a(2)*n(6,2)
            esc_6=a(3)*n(6,3)
            den_escalar=esc_4+esc_5+esc_6

            s(6)=num_escalar/den_escalar   !------- s is the distance between PI and P0

          !----------- Collision point against the plane 6 --------------------
            P_I(i,1)=i
            P_I(i,2) = P_0(i,2) + s(6)*a(1)
            P_I(i,3) = P_0(i,3) + s(6)*a(2)
            P_I(i,4) = P_0(i,4) + s(6)*a(3)
          !---------- PI - P0 = PIP0 (TO FIND THE COMPONENTS OF THE
COLLISION DISTANCE AGAINST THE PLANE 6)
            PIP0(i,1) = i
            PIP0(i,2) = (s(6)*a(1))
            PIP0(i,3) = (s(6)*a(2))
            PIP0(i,4) = (s(6)*a(3))
          end if
        end if
      end if
     end if
    end if
   end if

  end do

!----------------------------------------------------------------------------------------------------
! Calculating (P_I-P_0), in absolute value, for possible collision planes
! (Checking between all possible collision planes)
!----------------------------------------------------------------------------------------------------

  do i=1,6
    points_distance(i)=abs(s(i))   ! The distance between points P_0 and P_I is equal to
the absolute value of s(i)
  end do

!----------------------------------------------------------------------------------------------------
! Checking the shortest distance between points on the trajectory to the collision and
! determining the collision plane(s)
!----------------------------------------------------------------------------------------------------
```

```
stop_test="YES"     ! Assumpting there was no collision
minor_d=80000.
minor_d2=80000.
minor_d3=80000.
ncol=0
plane_col=0   !Assuming there was no collision against any plane
plane_col2=0   !There may be simultaneous collision between two planes
plane_col3=0   !There may be simultaneous collision between three planes (corner)

do i=1,6      ! In this loop, the collision planes are identified

  points_distance(i) = ANINT(points_distance(i)*100000.0)/100000.0    ! To ensure
accuracy in collision detection

   if ( (ncol.eq.0).and.(points_distance(i).lt.minor_d) ) then
    ncol=1
    plane_col=i
    minor_d=points_distance(i)

   else
    if (ncol.eq.1) then
      if (points_distance(i).eq.minor_d) then     ! particle collsion against two planes
simultaneously
         ncol=2
         plane_col2=i
         minor_d2=points_distance(i)
      end if

      if (points_distance(i).lt.minor_d) then
       ncol=1
       plane_col=i
       minor_d = points_distance(i)
      end if

    else
     if (ncol.eq.2) then

       if (points_distance(i).lt.minor_d2) then
         ncol=1
         plane_col=i
         minor_d= points_distance(i)
       else
        if (points_distance(i).eq.minor_d2) then    ! particle collsion against three
planes simultaneously
          ncol = 3
          plane_col3=i
          minor_d3=points_distance(i)
        end if
       end if
```

```
        end if
      end if
    end if
  end do

  if (ncol.ne.0) then
    ncol = 0
    if ((minor_d.lt.minor_d2).and.(minor_d.lt.minor_d3)) then
      ncol=ncol+1
      plane_col2=0
      plane_col3=0
    else
      if ((minor_d2.lt.minor_d).and.(minor_d2.lt.minor_d3)) then
        ncol=1
        plane_col=plane_col2
        minor_d=minor_d2
        plane_col2=0
        plane_col3=0
      else
        if ((minor_d3.lt.minor_d).and.(minor_d3.lt.minor_d2)) then
          ncol=1
          plane_col=plane_col3
          minor_d=minor_d3
          plane_col2=0
          plane_col3=0
        else               ! particle collsion against three planes simultaneously
          if ((minor_d.eq.minor_d2).and.(minor_d.eq.minor_d3)) then
            ncol=3
          else             ! particle collsion against two planes simultaneously
            if ((minor_d.eq.minor_d2).and.(minor_d.lt.minor_d3)) then
              ncol=2
              plane_col3=0
            else
              if ((minor_d.eq.minor_d3).and.(minor_d3.lt.minor_d2)) then
                ncol=2
                plane_col2=plane_col3
                plane_col3 = 0
              else
                if ((minor_d2.eq.minor_d3).and.(minor_d2.lt.minor_d)) then
                  ncol=2
                  plane_col=plane_col3
                  plane_col3 = 0
                end if
              end if
            end if
          end if
        end if
      end if
    end if
  end if
```

```
!-----------------------------------------------------------------------------------------------------
! Just checking the collision of the particle "n_part"
!-----------------------------------------------------------------------------------------------------
    if (ncol.ne.0) then  !!!E

     if (ncol.eq.1) then    !!!F         ! If there was a collision against a plane, the particle
position and velocity must be updated.
       stop_test="NO"

       if (plane_col.eq.1) then
        planecol="[1] - [RIGHT PLANE]"
       else
        if (plane_col.eq.2) then
         planecol="[2] - [LEFT PLANE]"
        else
         if (plane_col.eq.3) then
           planecol="[3] - [BOTTOM]"
         else
          if (plane_col.eq.4) then
            planecol="[4] - [UPPER PLANE]"
          else
            if (plane_col.eq.5) then
             planecol='[5] - [BACK PLANE]'
            else
              if (plane_col.eq.6) then
               planecol='[6] - [FRONTAL PLANE]'
              end if
            end if
          end if
         end if
        end if
       end if

!-----------------------------------------------------------------------------------------------------
! PERFORMING THE REFLECTION OF THE PARTICLE "n_part"
! Reflecting the final position of the center to the interior of the domain
!-----------------------------------------------------------------------------------------------------
      if (plane_col.eq.1) then    ! collision against the plane 01  [Right plane]

       if (CR.gt.0.) then
        reflected_position(1)   =   position_t1(n_part,2)+((1.0   +   CR)*   (radius-
distance(n_part,2)))*n(1,1)     !x
        reflected_position(2)              =              position_t1(n_part,3)+(2.*(radius-
distance(n_part,2)))*n(1,2)          !y
        reflected_position(3)              =              position_t1(n_part,4)+(2.*(radius-
distance(n_part,2)))*n(1,3)          !z
       else
        reflected_position(1)=P_I(1,2)-radius
        reflected_position(2)=P_I(1,3)
```

```
    reflected_position(3)=P_I(1,4)
   end if

  ! ---- Correcting particle velocity and updating velocity matrix after each collision
   velocity_t1(n_part,2)= (-1)*CR*velocity_t1(n_part,2)                    !x
   velocity_t1(n_part,3)= (1.-friction_coef)*velocity_t1(n_part,3)    !y
   velocity_t1(n_part,4)= (1.-friction_coef)*velocity_t1(n_part,4)    !z

  else
    if (plane_col.eq.2) then    ! collision against the plane 02  [Left plane]

      if (CR.gt.0.) then
        reflected_position(2)=position_t1(n_part,3)+                       (2.*(radius-
distance(n_part,3)))*n(2,2)              !y
        reflected_position(1)=position_t1(n_part,2)+((1.0       +          CR)*(radius-
distance(n_part,3)))*n(2,1)      !x
        reflected_position(3)=position_t1(n_part,4)+(2.*(radius-
distance(n_part,3)))*n(2,3)              !z

      else
        reflected_position(1)=P_I(2,2)+radius
        reflected_position(2)=P_I(2,3)
        reflected_position(3)=P_I(2,4)
      end if

  ! ---- Correcting particle velocity and updating velocity matrix after each collision
      velocity_t1(n_part,2)= (-1)*CR*velocity_t1(n_part,2)               !x
      velocity_t1(n_part,3)= (1.-friction_coef)*velocity_t1(n_part,3)   !y
      velocity_t1(n_part,4)= (1.-friction_coef)*velocity_t1(n_part,4)   !z

    else
      if (plane_col.eq.3) then    ! collision against the plane 03 [Bottom]
        if (CR.gt.0.) then
          reflected_position(1)=position_t1(n_part,2)+(2.*(radius-
distance(n_part,4)))*n(3,1)              !x
          reflected_position(2)=position_t1(n_part,3)+((1.0         +       CR)*(radius-
distance(n_part,4)))*n(3,2)  !y
          reflected_position(3)=position_t1(n_part,4)+(2.*(radius-
distance(n_part,4)))*n(3,3)              !z
        else
          reflected_position(1)=P_I(3,2)
          reflected_position(2)=P_I(3,3)+ radius
          reflected_position(3)=P_I(3,4)
        end if

  ! ---- Correcting particle velocity and updating velocity matrix after each collision
      velocity_t1(n_part,3)= (-1.)*CR*velocity_t1(n_part,3)                  !y
      velocity_t1(n_part,2)= (1.-friction_coef)*velocity_t1(n_part,2)      !x
      velocity_t1(n_part,4)= (1.-friction_coef)*velocity_t1(n_part,4)      !z
```

```
else
  if (plane_col.eq.4) then ! collision against the plane 04 [Upper Plane]

    if (CR.gt.0.) then
      reflected_position(1)=position_t1(n_part,2)+(2.*(radius-
distance(n_part,5)))*n(4,1)        !x
      reflected_position(2)=position_t1(n_part,3)+((1.0      +      CR)*(radius-
distance(n_part,5)))*n(4,2)        !y
      reflected_position(3)=position_t1(n_part,4)+(2.*(radius-
distance(n_part,5)))*n(4,3)        !z
    else
      reflected_position(1)=P_I(4,2)
      reflected_position(2)=P_I(4,3)-radius
      reflected_position(3)=P_I(4,4)
    end if

! ---- Correcting particle velocity and updating velocity matrix after each collision
    velocity_t1(n_part,3) = (-1)*CR*velocity_t1(n_part,3)              !y
    velocity_t1(n_part,2)= (1.-friction_coef)*velocity_t1(n_part,2)    !x
    velocity_t1(n_part,4)= (1.-friction_coef)*velocity_t1(n_part,4)    !z

  else
    if (plane_col.eq.5) then    ! collision against the plane 05 [Back Plane]

      if (CR.gt.0.) then
        reflected_position(3)=position_t1(n_part,4)+((1.0      +      CR)*(radius-
distance(n_part,6)))*n(5,3)  !z
        reflected_position(1)=position_t1(n_part,2)+(2.*(radius-
distance(n_part,6)))*n(5,1)        !x
        reflected_position(2)=position_t1(n_part,3)+(2.*(radius-
distance(n_part,6)))*n(5,2)        !y
      else
        reflected_position(1)=P_I(5,2)
        reflected_position(2)=P_I(5,3)
        reflected_position(3)=P_I(5,4)+radius
      end if

! ---- Correcting particle velocity and updating velocity matrix after each collision
      velocity_t1(n_part,4)= (-1)*CR*velocity_t1(n_part,4)              !z
      velocity_t1(n_part,2)= (1.-friction_coef)*velocity_t1(n_part,2)   !x
      velocity_t1(n_part,3) = (1.-friction_coef)*velocity_t1(n_part,3)  !y

    else
      if (plane_col.eq.6) then  ! collision against the plane 06 [Frontal Plane]

        if (CR.gt.0.) then
          reflected_position(3)=position_t1(n_part,4)+((1.0    +    CR)*(radius-
distance(n_part,7)))*n(6,3)  !z
          reflected_position(1)=position_t1(n_part,2)+(2.*(radius-
distance(n_part,7)))*n(6,1)        !x
```

```
                 reflected_position(2)=position_t1(n_part,3)+(2.*(radius-
distance(n_part,7)))*n(6,2)          !y
               else
                 reflected_position(1)=P_I(6,2)
                 reflected_position(2)=P_I(6,3)
                 reflected_position(3)=P_I(6,4)-radius
               end if

      ! ---- Correcting particle velocity and updating velocity matrix after each collision
               velocity_t1(n_part,4)= (-1)*CR*velocity_t1(n_part,4)          !z
               velocity_t1(n_part,2)= (1.-friction_coef)*velocity_t1(n_part,2)  !x
               velocity_t1(n_part,3)= (1.-friction_coef)*velocity_t1(n_part,3)  !y
              end if
           end if
          end if
        end if
       end if
      end if

!------------------------------------------------------------------------------------------------
! Obtaining the new position of the center of the particle "new C_0" for the new trajectory
! after the previous collisions
! This is the position of the mass center of the particle at the instant of the previous
! collision against the plane
!------------------------------------------------------------------------------------------------
      new_C0(1)=position_t0(n_part,2)+s(plane_col)*a(1)
      new_C0(2)=position_t0(n_part,3)+s(plane_col)*a(2)
      new_C0(3)=position_t0(n_part,4)+s(plane_col)*a(3)

!------------------------------------------------------------------------------------------------
! Updating the initial position matrix of the particle "n_part". This is the position of the
! particle center for the next iteration
!------------------------------------------------------------------------------------------------
      do j=1,3
       position_t0(n_part,j+1)=new_C0(j)
      end do

!------------------------------------------------------------------------------------------------
! Updating the matrix with the final position of the particle "n_part" !from the coordinates
! after reflection
! This is the final position of the particle mass centre after the !collision. If there are no
! more collisions, its position is already !defined.
!------------------------------------------------------------------------------------------------
      do j=1,3
        position_t1(n_part,j+1)=reflected_position(j)
      end do

!------------------------------------------------------------------------------------------------
!             COLLISION AGAINST TWO PLANES SIMULTANEOUSLY
!------------------------------------------------------------------------------------------------
```

```
else     !!!F
 if (ncol.eq.2) then
  stop_test="NO"

  if (plane_col.eq.1) then
   planecol="[1] - [RIGHT PLANE]"
  else
   if (plane_col.eq.2) then
    planecol="[2] - [LEFT PLANE]"
   else
    if (plane_col.eq.3) then
     planecol="[3] - [BOTTOM]"
    else
     if (plane_col.eq.4) then
      planecol="[4] - [UPPER]"
     else
      if (plane_col.eq.5) then
       planecol="[5] - [BACK PLANE]"
      else
       if (plane_col.eq.6) then
        planecol="[6] - [FRONTAL PLANE]"
       end if
      end if
     end if
    end if
   end if
  end if

  if (plane_col2.eq.1) then
   planecol="[1] - [RIGHT PLANE]"
  else
   if (plane_col2.eq.2) then
    planecol="[2] - [LEFT PLANE]"
   else
    if (plane_col2.eq.3) then
     planecol="[3] - [BOTTOM]"
    else
     if (plane_col2.eq.4) then
      planecol="[4] - [UPPER]"
     else
      if (plane_col2.eq.5) then
       planecol="[5] - [BACK PLANE]"
      else
       if (plane_col2.eq.6) then
        planecol="[6] - [FRONTAL PLANE]"
       end if
      end if
     end if
    end if
   end if
```

```
    end if

!---------------- First collision -------------------------------------------------------------------

    if (plane_col.eq.1) then  ! collision against the plane 01  [Right plane]

    if (CR.gt.0.) then
      reflected_position(1)  =  position_t1(n_part,2)+(  (1.0  +  CR)*  (radius-
distance(n_part,2)) )*n(1,1)    !x
      reflected_position(2)  =  position_t1(n_part,3)+(2.*(radius-
distance(n_part,2)))*n(1,2)    !y
      reflected_position(3)  =  position_t1(n_part,4)+(2.*(radius-
distance(n_part,2)))*n(1,3)    !z
     else
      reflected_position(1)=P_I(1,2)-radius
      reflected_position(2)=P_I(1,3)
      reflected_position(3)=P_I(1,4)
     end if

     !--- Correcting particle velocity and updating velocity matrix after each collision
      velocity_t1(n_part,2)= (-1)*CR*velocity_t1(n_part,2)          !x
      velocity_t1(n_part,3)= (1.-friction_coef)*velocity_t1(n_part,3)   !y
      velocity_t1(n_part,4)= (1.-friction_coef)*velocity_t1(n_part,4)   !z

    else
     if (plane_col.eq.2) then  ! collision against the plane 02  [Left plane]

      if (CR.gt.0.) then
       reflected_position(2)=position_t1(n_part,3)+  (2.*(radius-
distance(n_part,3)))*n(2,2)    !y
       reflected_position(1)=position_t1(n_part,2)+((1.0  +  CR)*(radius-
distance(n_part,3)))*n(2,1)    !x
       reflected_position(3)=position_t1(n_part,4)+(2.*(radius-
distance(n_part,3)))*n(2,3)    !z

      else
       reflected_position(1)=P_I(2,2)+radius
       reflected_position(2)=P_I(2,3)
       reflected_position(3)=P_I(2,4)
      end if

     !--- Correcting particle velocity and updating velocity matrix after each collision
      velocity_t1(n_part,2)= (-1)*CR*velocity_t1(n_part,2)          !x
      velocity_t1(n_part,3)= (1.-friction_coef)*velocity_t1(n_part,3)   !y
      velocity_t1(n_part,4)= (1.-friction_coef)*velocity_t1(n_part,4)   !z

     else
      if (plane_col.eq.3) then  ! collision against the plane 03  [Bottom]

       if (CR.gt.0.) then
```

```
            reflected_position(1)=position_t1(n_part,2)+(2.*(radius-
distance(n_part,4)))*n(3,1)        !x
            reflected_position(2)=position_t1(n_part,3)+((1.0        +        CR)*(radius-
distance(n_part,4)))*n(3,2) !y
            reflected_position(3)=position_t1(n_part,4)+(2.*(radius-
distance(n_part,4)))*n(3,3)        !z
          else
            reflected_position(1)=P_I(3,2)
            reflected_position(2)=P_I(3,3)+ radius
            reflected_position(3)=P_I(3,4)
          end if

      !--- Correcting particle velocity and updating velocity matrix after each collision
          velocity_t1(n_part,3)= (-1.)*CR*velocity_t1(n_part,3)                    !y
          velocity_t1(n_part,2)= (1.-friction_coef)*CR*velocity_t1(n_part,2)    !x
          velocity_t1(n_part,4)= (1.-friction_coef)*velocity_t1(n_part,4)         !z

        else
          if (plane_col.eq.4) then   ! collision against the plane 04 [Upper]

            if (CR.gt.0.) then
              reflected_position(1)=position_t1(n_part,2)+(2.*(radius-
distance(n_part,5)))*n(4,1)        !x
              reflected_position(2)=position_t1(n_part,3)+((1.0        +        CR)*(radius-
distance(n_part,5)))*n(4,2) !y
              reflected_position(3)=position_t1(n_part,4)+(2.*(radius-
distance(n_part,5)))*n(4,3)        !z
            else
              reflected_position(1)=P_I(4,2)
              reflected_position(2)=P_I(4,3)-radius
              reflected_position(3)=P_I(4,4)
            end if
      !--- Correcting particle velocity and updating velocity matrix after each collision
            velocity_t1(n_part,3) = (-1)*CR*velocity_t1(n_part,3)                  !y
            velocity_t1(n_part,2)= (1.-friction_coef)*velocity_t1(n_part,2)    !x
            velocity_t1(n_part,4)= (1.-friction_coef)*velocity_t1(n_part,4)    !z

          else
            if (plane_col.eq.5) then   ! collision against the plane 05 [Back Plane]

              if (CR.gt.0.) then
                reflected_position(3)=position_t1(n_part,4)+((1.0        +        CR)*(radius-
distance(n_part,6)))*n(5,3)        !z
                reflected_position(1)=position_t1(n_part,2)+(2.*(radius-
distance(n_part,6)))*n(5,1)        !x
                reflected_position(2)=position_t1(n_part,3)+(2.*(radius-
distance(n_part,6)))*n(5,2)        !y
              else
                reflected_position(1)=P_I(5,2)
                reflected_position(2)=P_I(5,3)
```

```
      reflected_position(3)=P_I(5,4)+radius
     end if

 !--- Correcting particle velocity and updating velocity matrix after each collision
     velocity_t1(n_part,4)= (-1)*CR*velocity_t1(n_part,4)                    !z
     velocity_t1(n_part,2)= (1.-friction_coef)*velocity_t1(n_part,2)    !x
     velocity_t1(n_part,3) = (1.-friction_coef)*velocity_t1(n_part,3)   !y

     else
      if (plane_col.eq.6) then  ! collision against the plane 06 [Frontal plane]

       if (CR.gt.0.) then
        reflected_position(3)=position_t1(n_part,4)+((1.0     +     CR)*(radius-
distance(n_part,7)))*n(6,3)  !z
        reflected_position(1)=position_t1(n_part,2)+(2.*(radius-
distance(n_part,7)))*n(6,1)           !x
        reflected_position(2)=position_t1(n_part,3)+(2.*(radius-
distance(n_part,7)))*n(6,2)           !y
       else
        reflected_position(1)=P_I(6,2)
        reflected_position(2)=P_I(6,3)
        reflected_position(3)=P_I(6,4)-radius
       end if

 !--- Correcting particle velocity and updating velocity matrix after each collision
       velocity_t1(n_part,4)= (-1)*CR*velocity_t1(n_part,4)                  !z
       velocity_t1(n_part,2)= (1.-friction_coef)*velocity_t1(n_part,2)  !x
       velocity_t1(n_part,3)= (1.-friction_coef)*velocity_t1(n_part,3)  !y

      end if
     end if
    end if
   end if
  end if
 end if

!-------------------------------------------------------------------------------------------
! Updating the final position matrix of the particle "n_part" from the coordinates after the
! first reflection
!-------------------------------------------------------------------------------------------

 do j=1,3
  position_t1(n_part,j+1)=reflected_position(j)
 end do

!--------------- Second collision ----------------------------------------------------------

 if (plane_col2.eq.1) then  ! collision against the plane 01 [Right plane]

  if (CR.gt.0.) then
```

```
        reflected_position(1) = position_t1(n_part,2)+( (1.0 + CR)* (radius-
distance(n_part,2)) )*n(1,1)          !x
        reflected_position(2) = position_t1(n_part,3)+(2.*(radius-
distance(n_part,2)))*n(1,2)          !y
        reflected_position(3) = position_t1(n_part,4)+(2.*(radius-
distance(n_part,2)))*n(1,3)          !z
      else
        reflected_position(1)=P_I(1,2)-radius
        reflected_position(2)=P_I(1,3)
        reflected_position(3)=P_I(1,4)
      end if

    !--- Correcting particle velocity and updating velocity matrix after each collision)
      velocity_t1(n_part,2)= (-1)*CR*velocity_t1(n_part,2)              !x
      velocity_t1(n_part,3)= (1.-friction_coef)*velocity_t1(n_part,3)   !y
      velocity_t1(n_part,4)= (1.-friction_coef)*velocity_t1(n_part,4)   !z

     else
      if (plane_col2.eq.2) then  ! collision against the plane 02 [Left plane]

        if (CR.gt.0.) then
          reflected_position(2)=position_t1(n_part,3)+ (2.*(radius-
distance(n_part,3)))*n(2,2)          !y
          reflected_position(1)=position_t1(n_part,2)+((1.0 + CR)*(radius-
distance(n_part,3)))*n(2,1)    !x
          reflected_position(3)=position_t1(n_part,4)+(2.*(radius-
distance(n_part,3)))*n(2,3)          !z

        else
          reflected_position(1)=P_I(2,2)+radius
          reflected_position(2)=P_I(2,3)
          reflected_position(3)=P_I(2,4)
        end if
    !--- Correcting particle velocity and updating velocity matrix after each collision)
        velocity_t1(n_part,2)= (-1)*CR*velocity_t1(n_part,2)              !x
        velocity_t1(n_part,3)= (1.-friction_coef)*velocity_t1(n_part,3)   !y
        velocity_t1(n_part,4)= (1.-friction_coef)*velocity_t1(n_part,4)   !z

      else
       if (plane_col2.eq.3) then  ! collision against the plane 03 [Bottom]

         if (CR.gt.0.) then
           reflected_position(1)=position_t1(n_part,2)+(2.*(radius-
distance(n_part,4)))*n(3,1)          !x
           reflected_position(2)=position_t1(n_part,3)+((1.0 + CR)*(radius-
distance(n_part,4)))*n(3,2)  !y
           reflected_position(3)=position_t1(n_part,4)+(2.*(radius-
distance(n_part,4)))*n(3,3)          !z
         else
           reflected_position(1)=P_I(3,2)
```

```
        reflected_position(2)=P_I(3,3)+ radius
        reflected_position(3)=P_I(3,4)
      end if

   !--- Correcting particle velocity and updating velocity matrix after each collision)
      velocity_t1(n_part,3)= (-1.)*CR*velocity_t1(n_part,3)                        !y
      velocity_t1(n_part,2)= (1.-friction_coef)*velocity_t1(n_part,2)       !x
      velocity_t1(n_part,4)= (1.-friction_coef)*velocity_t1(n_part,4)       !z

       else
        if (plane_col2.eq.4) then  ! collision against the plane 04  [Upper]

         if (CR.gt.0.) then
           reflected_position(1)=position_t1(n_part,2)+(2.*(radius-
distance(n_part,5)))*n(4,1)  !x
           reflected_position(2)=position_t1(n_part,3)+((1.0        +        CR)*(radius-
distance(n_part,5)))*n(4,2)  !y
           reflected_position(3)=position_t1(n_part,4)+(2.*(radius-
distance(n_part,5)))*n(4,3)  !z
         else
           reflected_position(1)=P_I(4,2)
           reflected_position(2)=P_I(4,3)-radius
           reflected_position(3)=P_I(4,4)
         end if

   !--- Correcting particle velocity and updating velocity matrix after each collision
        velocity_t1(n_part,3) = (-1)*CR*velocity_t1(n_part,3)             !y
        velocity_t1(n_part,2)= (1.-friction_coef)*velocity_t1(n_part,2)   !x
        velocity_t1(n_part,4)= (1.-friction_coef)*velocity_t1(n_part,4)   !z

        else
         if (plane_col2.eq.5) then   ! collision against the plane 05 [Back plane]

          if (CR.gt.0.) then
            reflected_position(3)=position_t1(n_part,4)+((1.0        +        CR)*(radius-
distance(n_part,6)))*n(5,3)          !z
            reflected_position(1)=position_t1(n_part,2)+(2.*(radius-
distance(n_part,6)))*n(5,1)          !x
            reflected_position(2)=position_t1(n_part,3)+(2.*(radius-
distance(n_part,6)))*n(5,2)          !y
          else
            reflected_position(1)=P_I(5,2)
            reflected_position(2)=P_I(5,3)
            reflected_position(3)=P_I(5,4)+radius
          end if

   !--- Correcting particle velocity and updating velocity matrix after each collision
          velocity_t1(n_part,4)= (-1)*CR*velocity_t1(n_part,4)                 !z
          velocity_t1(n_part,2)= (1.-friction_coef)*velocity_t1(n_part,2)    !x
          velocity_t1(n_part,3) = (1.-friction_coef)*velocity_t1(n_part,3)   !y
```

```
          else
           if (plane_col2.eq.6) then    ! collision against the plane 06 [Frontal plane]

            if (CR.gt.0.) then
             reflected_position(3)=position_t1(n_part,4)+((1.0      +      CR)*(radius-
distance(n_part,7)))*n(6,3)  !z
              reflected_position(1)=position_t1(n_part,2)+(2.*(radius-
distance(n_part,7)))*n(6,1)         !x
               reflected_position(2)=position_t1(n_part,3)+(2.*(radius-
distance(n_part,7)))*n(6,2)         !y
            else
             reflected_position(1)=P_I(6,2)
             reflected_position(2)=P_I(6,3)
             reflected_position(3)=P_I(6,4)-radius
            end if

!--- Correcting particle velocity and updating velocity matrix after each collision
            velocity_t1(n_part,4)= (-1)*CR*velocity_t1(n_part,4)             !z
            velocity_t1(n_part,2)= (1.-friction_coef)*velocity_t1(n_part,2)  !x
            velocity_t1(n_part,3)= (1.-friction_coef)*velocity_t1(n_part,3)  !y
           end if
         end if
        end if
       end if
      end if
     end if

!----------------------------------------------------------------------------------------------------
! Finding the position of the particle centre at the moment of collision
!----------------------------------------------------------------------------------------------------
     do i=1,3
      new_C0(i)= position_t0(n_part,i+1)
     end do

     if ((plane_col.eq.5).or.(plane_col.eq.6)) then
      new_C0(3)=position_t0(n_part,4)+s(plane_col)*a(3)
     else
      if ((plane_col.eq.1).or.(plane_col.eq.2)) then
       new_C0(1)=position_t0(n_part,2)+s(plane_col)*a(1)
      else
       if ((plane_col.eq.3).or.(plane_col.eq.4)) then
        new_C0(2)=position_t0(n_part,3)+s(plane_col)*a(2)
       end if
      end if
     end if

     if ((plane_col2.eq.5).or.(plane_col2.eq.6)) then
      new_C0(3)=position_t0(n_part,4)+s(plane_col2)*a(3)
     else
```

```
      if ((plane_col2.eq.1).or.(plane_col2.eq.2)) then
       new_C0(1)=position_t0(n_part,2)+s(plane_col2)*a(1)
      else
       if ((plane_col2.eq.3).or.(plane_col2.eq.4)) then
        new_C0(2)=position_t0(n_part,3)+s(plane_col2)*a(2)
       end if
      end if
     end if

!-------------------------------------------------------------------------------------
! Updating the initial position matrix of the particle "n_part". This is the position of the
! particle centre  for the next iteration
!-------------------------------------------------------------------------------------

     do j=1,3
      position_t0(n_part,j+1)=new_C0(j)
     end do

!-------------------------------------------------------------------------------------
! Updating the final position matrix of the particle "n_part" from the coordinates after the
! second reflection
!-------------------------------------------------------------------------------------

     do j=1,3
      position_t1(n_part,j+1)=reflected_position(j)
     end do

!-------------------------------------------------------------------------------------
! COLLISION AGAINST THREE PLANES SIMULTANEOUSLY
!-------------------------------------------------------------------------------------

     else   !!!F
      if (ncol.eq.3) then        ! IF THERE WAS A COLLISION AGAINST THREE
PLANES SIMULTANEOUSLY, THE PARTICLE POSITIONS AND VELOCITIES
MUST BE UPDATE
        stop_test="NO"

        if (plane_col.eq.1) then
         planecol="[1] - [RIGHT PLANE]"
        else
         if (plane_col.eq.2) then
          planecol="[2] - [LEFT PLANE]"
         else
          if (plane_col.eq.3) then
           planecol="[3] - [BOTTOM]"
            else
           if (plane_col.eq.4) then
            planecol="[4] - [UPPER]"
             else
            if (plane_col.eq.5) then
```

```
          planecol="[5] - [BACK PLANE]"
         else
          planecol="[6] - [FRONTAL PLANE]"
         end if
       end if
     end if
   end if
 end if

 if (plane_col2.eq.1) then
   planecol2="[1] - [RIGHT PLANE]"
 else
   if (plane_col2.eq.2) then
     planecol2="[2] - [LEFT PLANE]"
   else
     if (plane_col2.eq.3) then
       planecol2="[3] - [BOTTOM]"
         else
       if (plane_col2.eq.4) then
         planecol2="[4] - [UPPER]"
           else
         if (plane_col2.eq.5) then
           planecol2="[5] - [BACK PLANE]"
         else
           planecol2="[6] - [FRONTAL PLANE]"
         end if
       end if
     end if
   end if
 end if

 if (plane_col3.eq.1) then
   planecol3="[1] - [RIGHT PLANE]"
 else
   if (plane_col3.eq.2) then
     planecol3="[2] - [LEFT PLANE]"
   else
     if (plane_col3.eq.3) then
       planecol3="[3] - [BOTTOM]"
         else
       if (plane_col3.eq.4) then
         planecol3="[4] - [UPPER]"
           else
         if (plane_col3.eq.5) then
           planecol3="[5] - [BACK PLANE]"
         else
           planecol3="[6] - [FRONTAL PLANE]"
         end if
       end if
     end if
```

```
  end if
 end if


!---------------- First collision -----------------------------------------------------------------

 if (plane_col.eq.1) then    ! collision against the plane 01 [Right plane]

  if (CR.gt.0.) then
   reflected_position(1) = position_t1(n_part,2)+( (1.0 + CR)* (radius-
distance(n_part,2)) )*n(1,1)       !x
   reflected_position(2) = position_t1(n_part,3)+(2.*(radius-
distance(n_part,2)))*n(1,2)       !y
   reflected_position(3) = position_t1(n_part,4)+(2.*(radius-
distance(n_part,2)))*n(1,3)       !z
  else
   reflected_position(1)=P_I(1,2)-radius
   reflected_position(2)=P_I(1,3)
   reflected_position(3)=P_I(1,4)
  end if

!--- Correcting particle velocity and updating velocity matrix after each collision
  velocity_t1(n_part,2)= (-1)*CR*velocity_t1(n_part,2)              !x
  velocity_t1(n_part,3)= (1.-friction_coef)*velocity_t1(n_part,3)   !y
  velocity_t1(n_part,4)= (1.-friction_coef)*velocity_t1(n_part,4)   !z

 else
  if (plane_col.eq.2) then   ! collision against the plane 02 [Left plane]

   if (CR.gt.0.) then
    reflected_position(2)=position_t1(n_part,3)+ (2.*(radius-
distance(n_part,3)))*n(2,2)       !y
    reflected_position(1)=position_t1(n_part,2)+((1.0 + CR)*(radius-
distance(n_part,3)))*n(2,1)       !x
    reflected_position(3)=position_t1(n_part,4)+(2.*(radius-
distance(n_part,3)))*n(2,3)       !z

   else
    reflected_position(1)=P_I(2,2)+radius
    reflected_position(2)=P_I(2,3)
    reflected_position(3)=P_I(2,4)
   end if

!--- Correcting particle velocity and updating velocity matrix after each collision
   velocity_t1(n_part,2)= (-1)*CR*velocity_t1(n_part,2)              !x
   velocity_t1(n_part,3)= (1.-friction_coef)*velocity_t1(n_part,3)   !y
   velocity_t1(n_part,4)= (1.-friction_coef)*velocity_t1(n_part,4)   !z

  else
   if (plane_col.eq.3) then  ! collision against the plane 03 [Bottom]
```

```
            if (CR.gt.0.) then
              reflected_position(1)=position_t1(n_part,2)+(2.*(radius-
distance(n_part,4)))*n(3,1)        !x
              reflected_position(2)=position_t1(n_part,3)+((1.0     +     CR)*(radius-
distance(n_part,4)))*n(3,2)        !y
              reflected_position(3)=position_t1(n_part,4)+(2.*(radius-
distance(n_part,4)))*n(3,3)        !z
            else
             reflected_position(1)=P_I(3,2)
             reflected_position(2)=P_I(3,3)+ radius
             reflected_position(3)=P_I(3,4)
           end if

    !--- Correcting particle velocity and updating velocity matrix after each collision
           velocity_t1(n_part,3)= (-1.)*CR*velocity_t1(n_part,3)                !y
           velocity_t1(n_part,2)= (1.-friction_coef)*velocity_t1(n_part,2)      !x
           velocity_t1(n_part,4)= (1.-friction_coef)*velocity_t1(n_part,4)      !z

           else
             if (plane_col.eq.4) then   !  collision against the plane 04 [Upper]

              if (CR.gt.0.) then
               reflected_position(1)=position_t1(n_part,2)+(2.*(radius-
distance(n_part,5)))*n(4,1)        !x
               reflected_position(2)=position_t1(n_part,3)+((1.0     +     CR)*(radius-
distance(n_part,5)))*n(4,2)  !y
               reflected_position(3)=position_t1(n_part,4)+(2.*(radius-
distance(n_part,5)))*n(4,3)        !z
              else
               reflected_position(1)=P_I(4,2)
               reflected_position(2)=P_I(4,3)-radius
               reflected_position(3)=P_I(4,4)
              end if

    !--- Correcting particle velocity and updating velocity matrix after each collision
              velocity_t1(n_part,3) = (-1)*CR*velocity_t1(n_part,3)              !y
              velocity_t1(n_part,2)= (1.-friction_coef)*velocity_t1(n_part,2)   !x
              velocity_t1(n_part,4)= (1.-friction_coef)*velocity_t1(n_part,4)   !z

            else
              if (plane_col.eq.5) then    !  collision against the plane 05 [Back plane]

                if (CR.gt.0.) then
                 reflected_position(3)=position_t1(n_part,4)+((1.0     +     CR)*(radius-
distance(n_part,6)))*n(5,3)  !z
                 reflected_position(1)=position_t1(n_part,2)+(2.*(radius-
distance(n_part,6)))*n(5,1)  !x
                 reflected_position(2)=position_t1(n_part,3)+(2.*(radius-
distance(n_part,6)))*n(5,2)  !y
```

```
                else
                  reflected_position(1)=P_I(5,2)
                  reflected_position(2)=P_I(5,3)
                   reflected_position(3)=P_I(5,4)+radius
               end if

      !--- Correcting particle velocity and updating velocity matrix after each collision
               velocity_t1(n_part,4)= (-1)*CR*velocity_t1(n_part,4)              !z
               velocity_t1(n_part,2)= (1.-friction_coef)*velocity_t1(n_part,2)    !x
               velocity_t1(n_part,3)= (1.-friction_coef)*velocity_t1(n_part,3)    !y

              else
               if (plane_col.eq.6) then   !  collision against the plane 06  [Frontal plane]

               if (CR.gt.0.) then
                  reflected_position(3)=position_t1(n_part,4)+((1.0   +   CR)*(radius-
distance(n_part,7)))*n(6,3)   !z
                  reflected_position(1)=position_t1(n_part,2)+(2.*(radius-
distance(n_part,7)))*n(6,1)        !x
                  reflected_position(2)=position_t1(n_part,3)+(2.*(radius-
distance(n_part,7)))*n(6,2)        !y
                else
                  reflected_position(1)=P_I(6,2)
                  reflected_position(2)=P_I(6,3)
                  reflected_position(3)=P_I(6,4)-radius
                end if

      !--- Correcting particle velocity and updating velocity matrix after each collision
                velocity_t1(n_part,4)= (-1)*CR*velocity_t1(n_part,4)             !z
                velocity_t1(n_part,2)= (1.-friction_coef)*velocity_t1(n_part,2)  !x
                velocity_t1(n_part,3)= (1.-friction_coef)*velocity_t1(n_part,3)  !y

               end if
              end if
            end if
          end if
        end if
      end if

!---------------------------------------------------------------------------------------------
! Updating the final position matrix of the particle "n_part" from the coordinates after the
! first reflection
!---------------------------------------------------------------------------------------------

      do j=1,3
       position_t1(n_part,j+1)=reflected_position(j)
      end do

!---------------- Second collision --------------------------------------------------------------

      if (plane_col2.eq.1) then        ! colision with plane 01 [Right plane]
```

```
      if (CR.gt.0.) then
        reflected_position(1)  =  position_t1(n_part,2)+(  (1.0  +  CR)*  (radius-
distance(n_part,2)) )*n(1,1)           !x
        reflected_position(2)           =              position_t1(n_part,3)+(2.*(radius-
distance(n_part,2)))*n(1,2)            !y
        reflected_position(3)           =              position_t1(n_part,4)+(2.*(radius-
distance(n_part,2)))*n(1,3)            !z
      else
       reflected_position(1)=P_I(1,2)-radius
       reflected_position(2)=P_I(1,3)
       reflected_position(3)=P_I(1,4)
      end if

    !--- Correcting particle velocity and updating velocity matrix after each collision
      velocity_t1(n_part,2)= (-1)*CR*velocity_t1(n_part,2)                !x
      velocity_t1(n_part,3)= (1.-friction_coef)*velocity_t1(n_part,3)    !y
      velocity_t1(n_part,4)= (1.-friction_coef)*velocity_t1(n_part,4)    !z

     else
      if (plane_col2.eq.2) then    ! colision with plane 02 [Left plane]

       if (CR.gt.0.) then
         reflected_position(2)=position_t1(n_part,3)+                    (2.*(radius-
distance(n_part,3)))*n(2,2)          !y
         reflected_position(1)=position_t1(n_part,2)+((1.0          +          CR)*(radius-
distance(n_part,3)))*n(2,1)          !x
         reflected_position(3)=position_t1(n_part,4)+(2.*(radius-
distance(n_part,3)))*n(2,3)          !z
       else
         reflected_position(1)=P_I(2,2)+radius
         reflected_position(2)=P_I(2,3)
         reflected_position(3)=P_I(2,4)
       end if

    !--- Correcting particle velocity and updating velocity matrix after each collision
       velocity_t1(n_part,2)= (-1)*CR*velocity_t1(n_part,2)                !x
       velocity_t1(n_part,3)= (1.-friction_coef)*velocity_t1(n_part,3)    !y
       velocity_t1(n_part,4)= (1.-friction_coef)*velocity_t1(n_part,4)    !z

      else
       if (plane_col2.eq.3) then    ! colision with plane 03 [Bottom]

        if (CR.gt.0.) then
           reflected_position(1)=position_t1(n_part,2)+(2.*(radius-
distance(n_part,4)))*n(3,1)         !x
           reflected_position(2)=position_t1(n_part,3)+((1.0          +          CR)*(radius-
distance(n_part,4)))*n(3,2)         !y
         reflected_position(3)=position_t1(n_part,4)+(2.*(radius-
distance(n_part,4)))*n(3,3)         !z
        else
```

```
      reflected_position(1)=P_I(3,2)
      reflected_position(2)=P_I(3,3)+ radius
      reflected_position(3)=P_I(3,4)
     end if

!--- Correcting particle velocity and updating velocity matrix after each collision
     velocity_t1(n_part,3)= (-1.)*CR*velocity_t1(n_part,3)                    !y
     velocity_t1(n_part,2)= (1.-friction_coef)*velocity_t1(n_part,2)          !x
     velocity_t1(n_part,4)= (1.-friction_coef)*velocity_t1(n_part,4)          !z

     else
      if (plane_col2.eq.4) then           ! colision with plane 04 [Upper]

       if (CR.gt.0.) then
        reflected_position(1)=position_t1(n_part,2)+(2.*(radius-
distance(n_part,5)))*n(4,1)         !x
        reflected_position(2)=position_t1(n_part,3)+((1.0 + CR)*(radius-
distance(n_part,5)))*n(4,2)         !y
        reflected_position(3)=position_t1(n_part,4)+(2.*(radius-
distance(n_part,5)))*n(4,3)         !z
       else
        reflected_position(1)=P_I(4,2)
        reflected_position(2)=P_I(4,3)-radius
        reflected_position(3)=P_I(4,4)
       end if

!--- Correcting particle velocity and updating velocity matrix after each collision
       velocity_t1(n_part,3) = (-1)*CR*velocity_t1(n_part,3)                 !y
       velocity_t1(n_part,2)= (1.-friction_coef)*velocity_t1(n_part,2)    !x
       velocity_t1(n_part,4)= (1.-friction_coef)*velocity_t1(n_part,4)    !z

      else
       if (plane_col2.eq.5) then    ! colision with plane 05 [Back plane]

        if (CR.gt.0.) then
          reflected_position(3)=position_t1(n_part,4)+((1.0 + CR)*(radius-
distance(n_part,6)))*n(5,3)  !z
          reflected_position(1)=position_t1(n_part,2)+(2.*(radius-
distance(n_part,6)))*n(5,1)  !x
          reflected_position(2)=position_t1(n_part,3)+(2.*(radius-
distance(n_part,6)))*n(5,2)  !y
        else
          reflected_position(1)=P_I(5,2)
          reflected_position(2)=P_I(5,3)
          reflected_position(3)=P_I(5,4)+radius
        end if

!--- Correcting particle velocity and updating velocity matrix after each collision
        velocity_t1(n_part,4)= (-1)*CR*velocity_t1(n_part,4)                 !z
        velocity_t1(n_part,2)= (1.-friction_coef)*velocity_t1(n_part,2)    !x
```

```
            velocity_t1(n_part,3) = (1.-friction_coef)*velocity_t1(n_part,3)   !y

          else
            if (plane_col2.eq.6) then   !  colision with plane 06 [Frontal plane]

             if (CR.gt.0.) then
              reflected_position(3)=position_t1(n_part,4)+((1.0     +     CR)*(radius-
distance(n_part,7)))*n(6,3)       !z
               reflected_position(1)=position_t1(n_part,2)+(2.*(radius-
distance(n_part,7)))*n(6,1)        !x
               reflected_position(2)=position_t1(n_part,3)+(2.*(radius-
distance(n_part,7)))*n(6,2)        !y
             else
              reflected_position(1)=P_I(6,2)
              reflected_position(2)=P_I(6,3)
              reflected_position(3)=P_I(6,4)-radius
            end if

     !--- Correcting particle velocity and updating velocity matrix after each collision
            velocity_t1(n_part,4)= (-1)*CR*velocity_t1(n_part,4)               !z
            velocity_t1(n_part,2)= (1.-friction_coef)*velocity_t1(n_part,2)  !x
            velocity_t1(n_part,3)= (1.-friction_coef)*velocity_t1(n_part,3)  !y

           end if
          end if
         end if
        end if
       end if
      end if


!----------------------------------------------------------------------------------------------------
! Updating the final position matrix of the particle "n_part" from the coordinates after the
! first reflection
!----------------------------------------------------------------------------------------------------
      do j=1,3
       position_t1(n_part,j+1)=reflected_position(j)
      end do

!--------------- Third collision --------------------------------------------------------------

      if (plane_col3.eq.1) then    ! collision against the plane 01 [Right plane]

       if (CR.gt.0.) then
        reflected_position(1)   =   position_t1(n_part,2)+(   (1.0   +   CR)*   (radius-
distance(n_part,2)) )*n(1,1)        !x
        reflected_position(2)               =              position_t1(n_part,3)+(2.*(radius-
distance(n_part,2)))*n(1,2)        !y
        reflected_position(3)               =              position_t1(n_part,4)+(2.*(radius-
distance(n_part,2)))*n(1,3)        !z
```

```
    else
     reflected_position(1)=P_I(1,2)-radius
     reflected_position(2)=P_I(1,3)
     reflected_position(3)=P_I(1,4)
    end if

  !--- Correcting particle velocity and updating velocity matrix after each collision
    velocity_t1(n_part,2)= (-1)*CR*velocity_t1(n_part,2)                !x
    velocity_t1(n_part,3)= (1.-friction_coef)*velocity_t1(n_part,3)     !y
    velocity_t1(n_part,4)= (1.-friction_coef)*velocity_t1(n_part,4)     !z

   else
    if (plane_col3.eq.2) then   ! collision against the plane 02 [Left plane]

     if (CR.gt.0.) then
       reflected_position(2)=position_t1(n_part,3)+        (2.*(radius-
distance(n_part,3)))*n(2,2)          !y
       reflected_position(1)=position_t1(n_part,2)+((1.0      +      CR)*(radius-
distance(n_part,3)))*n(2,1)     !x
       reflected_position(3)=position_t1(n_part,4)+(2.*(radius-
distance(n_part,3)))*n(2,3)          !z

     else
       reflected_position(1)=P_I(2,2)+radius
       reflected_position(2)=P_I(2,3)
       reflected_position(3)=P_I(2,4)
     end if

  !--- Correcting particle velocity and updating velocity matrix after each collision
     velocity_t1(n_part,2)= (-1)*CR*velocity_t1(n_part,2)                !x
     velocity_t1(n_part,3)= (1.-friction_coef)*velocity_t1(n_part,3)     !y
     velocity_t1(n_part,4)= (1.-friction_coef)*velocity_t1(n_part,4)     !z

    else
     if (plane_col3.eq.3) then  ! collision against the plane 03  [Bottom]

      if (CR.gt.0.) then
        reflected_position(1)=position_t1(n_part,2)+(2.*(radius-
distance(n_part,4)))*n(3,1)          !x
        reflected_position(2)=position_t1(n_part,3)+((1.0      +      CR)*(radius-
distance(n_part,4)))*n(3,2)  !y
        reflected_position(3)=position_t1(n_part,4)+(2.*(radius-
distance(n_part,4)))*n(3,3)          !z
      else
        reflected_position(1)=P_I(3,2)
        reflected_position(2)=P_I(3,3)+ radius
        reflected_position(3)=P_I(3,4)
      end if

  !--- Correcting particle velocity and updating velocity matrix after each collision
```

```
        velocity_t1(n_part,3)= (-1.)*CR*velocity_t1(n_part,3)                    !y
        velocity_t1(n_part,2)= (1.-friction_coef)*velocity_t1(n_part,2)      !x
        velocity_t1(n_part,4)= (1.-friction_coef)*velocity_t1(n_part,4)      !z

        else
         if (plane_col3.eq.4) then  ! collision against the plane 04  [Upper]

          if (CR.gt.0.) then
            reflected_position(1)=position_t1(n_part,2)+(2.*(radius-
distance(n_part,5)))*n(4,1)        !x
            reflected_position(2)=position_t1(n_part,3)+((1.0        +        CR)*(radius-
distance(n_part,5)))*n(4,2)  !y
            reflected_position(3)=position_t1(n_part,4)+(2.*(radius-
distance(n_part,5)))*n(4,3)        !z
          else
           reflected_position(1)=P_I(4,2)
           reflected_position(2)=P_I(4,3)-radius
           reflected_position(3)=P_I(4,4)
          end if

    !--- Correcting particle velocity and updating velocity matrix after each collision
          velocity_t1(n_part,3) = (-1)*CR*velocity_t1(n_part,3)                  !y
          velocity_t1(n_part,2)= (1.-friction_coef)*velocity_t1(n_part,2)      !x
          velocity_t1(n_part,4)= (1.-friction_coef)*velocity_t1(n_part,3)      !z

         else
          if (plane_col3.eq.5) then    ! collision against the plane 05 [Back plane]

           if (CR.gt.0.) then
            reflected_position(3)=position_t1(n_part,4)+((1.0        +        CR)*(radius-
distance(n_part,6)))*n(5,3)  !z
            reflected_position(1)=position_t1(n_part,2)+(2.*(radius-
distance(n_part,6)))*n(5,1)        !x
            reflected_position(2)=position_t1(n_part,3)+(2.*(radius-
distance(n_part,6)))*n(5,2)        !y
           else
            reflected_position(1)=P_I(5,2)
            reflected_position(2)=P_I(5,3)
            reflected_position(3)=P_I(5,4)+radius
          end if

    !--- Correcting particle velocity and updating velocity matrix after each collision
          velocity_t1(n_part,4)= (-1)*CR*velocity_t1(n_part,4)                  !z
          velocity_t1(n_part,2)= (1.-friction_coef)*velocity_t1(n_part,2)    !x
          velocity_t1(n_part,3) = (1.-friction_coef)*velocity_t1(n_part,3)   !y

          else
           if (plane_col3.eq.6) then    ! collision against the plane 06 [Frontal plane]

            if (CR.gt.0.) then
```

```
              reflected_position(3)=position_t1(n_part,4)+((1.0      +      CR)*(radius-
distance(n_part,7)))*n(6,3)         !z
              reflected_position(1)=position_t1(n_part,2)+(2.*(radius-
distance(n_part,7)))*n(6,1)         !x
              reflected_position(2)=position_t1(n_part,3)+(2.*(radius-
distance(n_part,7)))*n(6,2)         !y
             else
              reflected_position(1)=P_I(6,2)
              reflected_position(2)=P_I(6,3)
              reflected_position(3)=P_I(6,4)-radius
             end if

    !--- Correcting particle velocity and updating velocity matrix after each collision
             velocity_t1(n_part,4)= (-1)*CR*velocity_t1(n_part,4)             !z
             velocity_t1(n_part,2)= (1.-friction_coef)*velocity_t1(n_part,2) !x
             velocity_t1(n_part,3)= (1.-friction_coef)*velocity_t1(n_part,3) !y
            end if
          end if
        end if
      end if
    end if
  end if

!------------------------------------------------------------------------------------------
! DETERMINING THE POSITION OF THE PARTICLE CENTER AT THE MOMENT
! OF COLLISION
!------------------------------------------------------------------------------------------

  if ((plane_col.eq.5).or.(plane_col.eq.6)) then          ! collision against first plane
   new_C0(3)=position_t0(n_part,4)+s(plane_col)*a(3)
  else
   if ((plane_col.eq.1).or.(plane_col.eq.2)) then
    new_C0(1)=position_t0(n_part,2)+s(plane_col)*a(1)
   else
    if ((plane_col.eq.3).or.(plane_col.eq.4)) then
     new_C0(2)=position_t0(n_part,3)+s(plane_col)*a(2)
    end if
   end if
  end if

  if ((plane_col2.eq.5).or.(plane_col2.eq.6)) then          ! collision against second plane
   new_C0(3)=position_t0(n_part,4)+s(plane_col2)*a(3)
  else
   if ((plane_col2.eq.1).or.(plane_col2.eq.2)) then
    new_C0(1)=position_t0(n_part,2)+s(plane_col2)*a(1)
   else
    if ((plane_col2.eq.3).or.(plane_col2.eq.4)) then
     new_C0(2)=position_t0(n_part,3)+s(plane_col2)*a(2)
    end if
   end if
```

```
      end if

      if ((plane_col3.eq.5).or.(plane_col3.eq.6)) then        ! collision against third plane
       new_C0(3)=position_t0(n_part,4)+s(plane_col3)*a(3)
      else
       if ((plane_col3.eq.1).or.(plane_col3.eq.2)) then
        new_C0(1)=position_t0(n_part,2)+s(plane_col3)*a(1)
       else
        if ((plane_col3.eq.3).or.(plane_col3.eq.4)) then
         new_C0(2)=position_t0(n_part,3)+s(plane_col3)*a(2)
        end if
       end if
      end if

!------------------------------------------------------------------------------------------------
! Updating the initial position matrix of the particle "n_part". This is the position of the
! particle center for the next iteration
!------------------------------------------------------------------------------------------------
      do j=1,3
       position_t0(n_part,j+1)=new_C0(j)
      end do

!------------------------------------------------------------------------------------------------
! Updating the final position of the particle "n_part" from the coordinates after reflection
! This is the final position of the particle after the collision. If theere are more collisions,
! its position is already defined
!------------------------------------------------------------------------------------------------
      do j=1,3
        position_t1(n_part,j+1)=reflected_position(j)
      end do
     end if  !!! collision against 3 planes simultaneously
     end if  !!! collision against 2 planes simultaneously

   end if   !!! collision against 1 plane

  end if  !!!E

 else  !!!!B   ! It will only enter this ELSE if there is no more collision (distances to planes
greater than zero)

    stop_test="YES"

 end if  !!!!B

end do !!!AAA

stop_test="NO"

!------------------------------------------------------------------------------------------------
```

```
!Great! All collisions for the particle "n_part"  have been performed at the current time step.
!Now, let's change the particle
!-----------------------------------------------------------------------------------------------

end do  ! Particle change

rewind(46)
rewind(45)

do i=1,ntotal
  write(46,54) int(position_t1(i,1)), (position_t1(i,j),j=2,4)
  write(45,54) int(velocity_t1(i,1)), (velocity_t1(i,j),j=2,4)
end do
54  format(I8, 3(1x,D14.7))
55  format(3(1x,D14.7))

do i=1,ntotal
  x(i,1) = position_t1(i,2)
  x(i,2) = position_t1(i,3)
  x(i,3) = position_t1(i,4)

  v(i,1) = velocity_t1(i,2)
  v(i,2) = velocity_t1(i,3)
  v(i,3) = velocity_t1(i,4)
end do

end subroutine collision_test


subroutine output(ntotal,itimestep,recording_step,p,hsml)

!------------------------------------------------------------------------------------------------
!   Subroutine for saving particle information to output files
!
!  ntotal --  Total particle number                                                        [in]
!  itimestep -- Current iteration                                                          [in]
!  hsml --   Smoothing length                                                              [in]
!------------------------------------------------------------------------------------------------

   implicit none
   include 'param.inc'

   integer ntotal, aux, i, d, itype(maxn), funit, funit2, funit3, funit4, itimestep, recording_step
   double precision xpos(maxn,4), auxv(maxn,4), x(maxn,dim), v(maxn,dim), p(maxn), hsml(maxn)
   character(len=25) aux2
   character(len=45) name_pos, name_vel, name_pres, name_hsml
```

```
rewind(45)
rewind(46)
do i=1, ntotal
 read(46,*) (xpos(i,d), d=1,4)
 read(45,*) (auxv(i,d), d=1,4)
end do

aux=itimestep
if ((itimestep.eq.1).or.(mod(itimestep,recording_step).eq.0)) then

 write(aux2,*) itimestep            ! Converting an integer to a character

 if (itimestep.eq.1) then
  funit=itimestep+3100
  funit2=itimestep+60100
  funit3=itimestep+91000
  funit4=itimestep+121000
 else
  funit=int(itimestep/recording_step)+3100
  funit2=int(itimestep/recording_step)+61000
  funit3=int(itimestep/recording_step)+91000
  funit4=int(itimestep/recording_step)+121000
 end if

 if (aux.lt.10) name_pos='../output/data/position/000000'//trim(adjustl(aux2))//'.dat'
 if ((aux.ge.10).and.(aux.lt.100)) name_pos='../output/data/position/00000'//trim(adjustl(aux2))//'.dat'
 if ((aux.ge.100).and.(aux.lt.1000)) name_pos='../output/data/position/0000'//trim(adjustl(aux2))//'.dat'
 if ((aux.ge.1000).and.(aux.lt.10000)) name_pos='../output/data/position/000'//trim(adjustl(aux2))//'.dat'
 if ((aux.ge.10000).and.(aux.lt.100000)) name_pos='../output/data/position/00'//trim(adjustl(aux2))//'.dat'
 if ((aux.ge.100000).and.(aux.lt.1000000)) name_pos='../output/data/position/0'//trim(adjustl(aux2))//'.dat'
 if (aux.eq.1000000) name_pos='../output/data/position'//trim(adjustl(aux2))//'.dat'

 open(unit=funit,file=name_pos)
 do i = 1, ntotal
  write (funit,221) i, xpos(i,2), xpos(i,3), xpos(i,4)
 enddo

 if (aux.lt.10) name_vel='../output/data/velocity/000000'//trim(adjustl(aux2))//'.dat'
 if ((aux.ge.10).and.(aux.lt.100)) name_vel='../output/data/velocity/00000'//trim(adjustl(aux2))//'.dat'
 if ((aux.ge.100).and.(aux.lt.1000)) name_vel='../output/data/velocity/0000'//trim(adjustl(aux2))//'.dat'
 if ((aux.ge.1000).and.(aux.lt.10000)) name_vel='../output/data/velocity/000'//trim(adjustl(aux2))//'.dat'
```

```
    if ((aux.ge.10000).and.(aux.lt.100000))
name_vel='../output/data/velocity/00'//trim(adjustl(aux2))//'.dat'
    if ((aux.ge.100000).and.(aux.lt.1000000))
name_vel='../output/data/velocity/0'//trim(adjustl(aux2))//'.dat'
    if (aux.eq.1000000) name_vel='../output/data/velocity'//trim(adjustl(aux2))//'.dat'

    open(unit=funit2,file=name_vel)
    do i = 1, ntotal
      write (funit2,221) i, auxv(i,2), auxv(i,3), auxv(i,4)
    end do

    if (aux.lt.10)  name_pres='../output/data/pressure/000000'//trim(adjustl(aux2))//'.dat'
    if ((aux.ge.10).and.(aux.lt.100))
name_pres='../output/data/pressure/00000'//trim(adjustl(aux2))//'.dat'
    if ((aux.ge.100).and.(aux.lt.1000))
name_pres='../output/data/pressure/0000'//trim(adjustl(aux2))//'.dat'
    if ((aux.ge.1000).and.(aux.lt.10000))
name_pres='../output/data/pressure/000'//trim(adjustl(aux2))//'.dat'
    if ((aux.ge.10000).and.(aux.lt.100000))
name_pres='../output/data/pressure/00'//trim(adjustl(aux2))//'.dat'
    if ((aux.ge.100000).and.(aux.lt.1000000))
name_pres='../output/data/pressure/0'//trim(adjustl(aux2))//'.dat'
    if (aux.eq.1000000) name_pres='../output/data/pressure'//trim(adjustl(aux2))//'.dat'

    open(unit=funit3,file=name_pres)

    do i = 1, ntotal
      write (funit3,222) i, xpos(i,2), xpos(i,3), xpos(i,4), p(i)
    end do

    if (aux.lt.10)  name_hsml='../output/data/hsml/000000'//trim(adjustl(aux2))//'.dat'
    if ((aux.ge.10).and.(aux.lt.100))
name_hsml='../output/data/hsml/00000'//trim(adjustl(aux2))//'.dat'
    if ((aux.ge.100).and.(aux.lt.1000))
name_hsml='../output/data/hsml/0000'//trim(adjustl(aux2))//'.dat'
    if ((aux.ge.1000).and.(aux.lt.10000))
name_hsml='../output/data/hsml/000'//trim(adjustl(aux2))//'.dat'
    if ((aux.ge.10000).and.(aux.lt.100000))
name_hsml='../output/data/hsml/00'//trim(adjustl(aux2))//'.dat'
    if ((aux.ge.100000).and.(aux.lt.1000000))
name_hsml='../output/data/hsml/0'//trim(adjustl(aux2))//'.dat'
    if (aux.eq.1000000) name_hsml='../output/data/hsml'//trim(adjustl(aux2))//'.dat'

    open(unit=funit4,file=name_hsml)

    do i = 1, ntotal
      write (funit4,223) i, hsml(i)
    end do

 221  format (I8, 3(1x,D21.14))
```

```
 222  format (I8, 4(1x,D21.14))
 223  format(I8, 1x, D21.14)

      close(funit)
      close(funit2)
      close(funit3)
      close(funit4)

    end if

end subroutine output
```

```
!-----------------------------------------------------------------------------------------------
! Here are the parameters and constants used in the SPH software
! A file must be created with the name param.inc
!-----------------------------------------------------------------------------------------------

!    dim : Dimemsion of the problem (1, 2 or 3)
     integer dim
     parameter (dim = 3)

!    maxn     : Maximum number of particles
     integer maxn, n_neighbours
     parameter (maxn = 50000, n_neighbours = 150)

     double precision pi
     parameter ( pi = 3.14159265358979323846 )

!     alpha coefficient (in artificial viscosity calculation)
      double precision  alpha
      parameter ( alpha = 0.05 )

!----------------------------------------------------------------------------------
!     Artificial Pressure
      character (len = 3) Artificial_press
      parameter (Artificial_press = 'YES')

      double precision lambda
      parameter (lambda = 0.10)
!----------------------------------------------------------------------------------

!     XSPH Correction
      character (len = 3) XSPH_correction
      parameter (XSPH_correction = 'NO')
```

```
# ============================================================
# makefile
#============================================================
/folder_address_of_item_1/fonts/main.run: main.o input.o geometry.o h_upgrade.o
direct_find.o kernel.o continuity_density.o pressure.o \
internal_force.o external_force.o acceleration.o XSPH.o time_integration.o
free_surface.o collision_test.o output.o \

#
	ifort -O3 -o /folder_address_of_item_1/fonts/main.run main.o input.o geometry.o
h_upgrade.o direct_find.o kernel.o continuity_density.o pressure.o \
internal_force.o external_force.o acceleration.o XSPH.o time_integration.o
free_surface.o collision_test.o output.o \
#
#

main.o:	param.inc main.f90
	ifort -O3 -c main.f90
input.o: param.inc input.f90
	ifort -O3 -c input.f90
geometry.o: param.inc geometry.f90
	ifort -O3 -c geometry.f90
h_upgrade.o: param.inc h_upgrade.f90
	ifort -O3 -c h_upgrade.f90
direct_find.o: param.inc direct_find.f90
	ifort -O3 -c direct_find.f90
kernel.o:	param.inc kernel.f90
	ifort -O3 -c kernel.f90
continuity_density.o: param.inc continuity_density.f90
	ifort -O3 -c continuity_density.f90
pressure.o: param.inc pressure.f90
	ifort -O3 -c pressure.f90
internal_force.o: param.inc internal_force.f90
	ifort -O3 -c internal_force.f90
external_force.o: param.inc external_force.f90
	ifort -O3 -c external_force.f90
acceleration.o: param.inc acceleration.f90
	ifort -O3 -c acceleration.f90
XSPH.o:	param.inc XSPH.f90
	ifort -O3 -c XSPH.f90
time_integration.o: param.inc time_integration.f90
	ifort -O3 -c time_integration.f90
free_surface.o:	param.inc free_surface.f90
	ifort -O3 -c free_surface.f90
collision_test.o: param.inc collision_test.f90
	ifort -O3 -c collision_test.f90
output.o:	param.inc output.f90
	ifort -O3 -c output.f90
# ============================================================
# clean
```

```
#=================================================================
clean:
	rm -rf *.o
```

Guidelines for implementing the program:

1. Create individual files (*.f90) for the *main program* and *subroutines*, besides *param.inc* and *Makefile*
2. Create a folder with the chosen name to paste all files
3. Inside the previous folder, create the *fonts*, *output* and *figures* folders
4. Inside *output*, create the *data* folder with the folders named *acceleration*, *density*, *free_surface, hsml*, *position*, *pressure*, *pressure_force*, *velocity* and *viscous_force*
5. Create a file named *Makefile* with the content presented above
6. Set the folder address of item 2 in the *Makefile*
7. Open the terminal inside */folder_address_of item_2/fonts/* and give the command make
8. Run the program with the command ./main.run

Note: Inside the output folder there is a script to generate graphs showing the evolution of fluid particles over time.

Conclusions

As a final consideration, this author would like to leave these few words to readers to clarify that this book seeks to make its scientific contribution by providing a routine that treats boundaries reflexively in particle simulation methods. The routine for treating boundaries is completely independent of the particle method employed to solve the physical conservation equations. In this way, the applications of reflexive boundary conditions (presented here in conjunction with the SPH method) can be extended to other particle methods.

Over more than 10 years, these boundary conditions have been applied and tested in some case studies in two-dimensional and three-dimensional domains. Through the publication of the scientific papers listed in the introductory section of this book and some more - available on the research dissemination website https://carlosdutrafraga.com - this author concludes his current contribution to the scientific community through this simple work, which aims to make public and with the potential to be contributory to some interested researchers the development and implementation of reflective physical boundary conditions.

C. A. Dutra Fraga Filho, *Computational Algorithms for Reflective Boundary Conditions in SPH Fluid Dynamics Simulation*, Synthesis Lectures on Mechanical Engineering,
https://doi.org/10.1007/978-3-031-99370-1

The manufacturer's authorised representative in the EU is Springer
Nature Customer Service Centre GmbH, Europaplatz 3, 69115 Heidelberg,
Germany. If you have any concerns regarding our products, please
contact ProductSafety@springernature.com

Printed and bound by CPI Group (UK) Ltd, Croydon, CR0 4YY
07/08/2026
02189671-0004